Brooks/Cole

One-Unit Series

in Precalculus Mathematics

Introduction To

SYMBOLIC LOGIC

Second Edition

Brooks/Cole

One-Unit Series

in Precalculus Mathematics

INTRODUCTION TO

SYMBOLIC LOGIC

Second Edition

Karl J. Smith

BROOKS/COLE PUBLISHING COMPANY

Pacific Grove, California

Brooks/Cole Publishing Company
A Division of Wadsworth, Inc.

Printed in the United States of America

10 9 8 7 6 5 4 3 2 1

Library of Congress Cataloging-in-Publication Data

Smith, Karl J.
Introduction to Symbolic Logic/ Karl J. Smith -- 2nd ed.
p. cm. -- (Brooks/Cole one-unit series in precalculus mathematics)
Includes index.
ISBN 0-534-14931-6 (paper)
1. Logic. Symbolic and mathematical I. Title II. Series
BC135.S62 1991
511--dc20 90-24933
CIP

Sponsoring Editor: Paula-Christy Heighton
Editorial Assistant: Elaine Giuliano
Marketing Representative: Deborah Mobley
Production Editor: Joan Marsh
Production Service: Bookman Productions
Manuscript Editor: Steven Gray
Interior Design: Karl J. Smith
Cover Design: Vernon T. Boes
Typesetting: Karlin Enterprises
Cover Printing: Malloy Lithographing, Inc.
Printing and Binding: Malloy Lithographing, Inc.

PREFACE

As its title suggests, this book is designed to provide an introduction to symbolic logic. Since almost no previous knowledge of the subject is assumed, the book can be used as a supplementary text in a wide variety of courses. It can also be used as the basis for a short (one-unit) course, or it can be studied independently, which is why the answers to all the problems are provided in an appendix.

The primary goal of this book is to provide a foundation for evaluating logical arguments—not only in a formal statement–proof context in mathematics, but also in everyday discussions. Examples are taken from mathematics, newspapers, radio and television commercials, and other contemporary sources in an effort to give the reader a course with real utility. By studying symbolic logic, you can sharpen your perception of validity, your ability to recognize fallacies, and your ability to evaluate lengthy arguments.

Chapter 10 assumes a knowledge of the closure, commutative, associative, identity, inverse, and distributive properties, as well as of sets and set operations.

I would like to extend my thanks to the following reviewers for their comments and suggestions: William Ash, SUNY-Tech College; Carl Hall, University of Texas, El Paso; Rick Patrick, Adirondack Community College; and Marcella Laddon, Monterey Peninsula College.

In particular, my appreciation to Carol Kublin for checking all of the examples and answers for accuracy.

Karl J. Smith
Sebastopol, CA

CONTENTS

Preface v

CHAPTER 1 **Fundamentals** 1

1.1 What Is Symbolic Logic? 1
1.2 Statements 2
1.3 Conjunction 3
1.4 Disjunction 4
1.5 Negation 6
1.6 Translating and Combining Connectives 8
1.7 Problem Set 1 9

CHAPTER 2 **Truth Tables** 11

2.1 Definition and Examples 11
2.2 Problem Set 2 13

CHAPTER 3 **The Conditional** 15

3.1 If–Then 16
3.2 Converse, Inverse, and Contrapositive 18
3.3 Problem Set 3 20

CHAPTER 4 **Additional Operators** 23

4.1 Biconditional 23
4.2 Miscellaneous Operators 24
4.3 Problem Set 4 25

CHAPTER 5 **Tautologies** 28

5.1 Definition 28
5.2 Implication 29
5.3 Laws of Logic 29
5.4 The Trivial Tautology, $p \Leftrightarrow p$ 30
5.5 The Law of Double Negation, $p \Leftrightarrow \sim(\sim p)$ 30
5.6 The Law of the Excluded Middle, $p \vee (\sim p)$ 30
5.7 Direct Reasoning, $[(p \rightarrow q) \wedge p] \Rightarrow q$ 30
5.8 Indirect Reasoning, $[(p \rightarrow q) \wedge (\sim q)] \Rightarrow (\sim p)$ 31
5.9 Law of Transitivity, $[(p \rightarrow q) \wedge (q \rightarrow r)] \Rightarrow (p \rightarrow r)$ 32
5.10 Law of Contraposition, $(p \rightarrow q) \Leftrightarrow (\sim q \rightarrow \sim p)$ 33
5.11 Disjunctive Syllogism, $[(p \vee q) \wedge (\sim p)] \Rightarrow q$ 33
5.12 Problem Set 5 34

CHAPTER 6 **Logical Fallacies** 36

6.1 Validity 36
6.2 Fallacy of Assuming the Consequent 36

6.3 Fallacy of Denying the Antecedent 37
6.4 False Chain Pattern 38
6.5 Problem Set 6 38

CHAPTER 7 **Quantifiers and Euler Circles** 41

7.1 Universal and Existential Quantifiers 41
7.2 Truth Values of Quantified Sentences 42
7.3 Introduction to Euler Circles 43
7.4 Simple Connectives 44
7.5 Logical Arguments 46
7.6 Problem Set 7 48

CHAPTER 8 **The Nature of Proof** 50

8.1 What Is Proof? 50
8.2 Examples of Logical Arguments 50
8.3 Problem Set 8 52

*CHAPTER 9 **Puzzle Problems in Logic** 54

9.1 An Alternative Method of Proof 54
9.2 Problem Set 9 58

*CHAPTER 10 **Logic Circuits** 60

10.1 Schematics 60
10.2 Negation 61
10.3 Conjunction 62
10.4 Disjunction 62
10.5 Logical Gates 63
10.6 Problem Set 10 65

*CHAPTER 11 **Analysis of Statements** 67

11.1 Mathematical Systems 67
11.2 Closure Property 67
11.3 Commutative Property 68
11.4 Associative Property 68
11.5 Distributive Property 68
11.6 Identity Elements 69
11.7 Inverse and Property of Complementation 69
11.8 Boolean Algebra 70
11.9 Boolean Algebra of Logic, Circuits, and Sets 70
11.10 Problem Set 11 72

CHAPTER 12 **Summary and Review** 73

12.1 Outline 73
12.2 Problem Set 12 75

APPENDIX **Answers** 78

Index 95

*Optional chapters

CHAPTER 1 FUNDAMENTALS

1.1 What Is Symbolic Logic?

We are familiar with logic and with the idea that some people have "logical minds" whereas others do not. How, then, can we all become logical? It is not always easy to follow lengthy arguments in order to come up with valid conclusions. The purpose of symbolic logic is to set up an artificial symbolic language that can be used to simplify complicated logical arguments. The great German mathematician Gottfried Leibniz (1646–1716) was the first to think of this approach when at the age of 14 he attempted to reform classical logic. He called symbolic logic the *universal characteristic* and wrote in 1666 that he wanted to create

> a general method in which all truths of the reason would be reduced to a kind of calculation. At the same time this would be a sort of universal language or script, but infinitely different from all those projected hitherto; for the symbols and even the words in it would direct the reason; and errors, except those of fact, would be mere mistakes in calculation. It would be very difficult to form or invent this language or characteristic, but very easy to understand it without any dictionaries.

This dream was not realized until the English mathematician George Boole (1815–1864) separated the symbols of mathematical operations from the things upon which they operated and set up a simple, workable system of symbolic logic. In 1859 he set forth his ideas in *An Investigation of the Laws of Thought.* Unfortunately, this work was not well received, and it wasn't until Bertrand Russell (1872–1970) and Alfred North Whitehead (1861–1947) used symbolic logic in their *Principia Mathematica* that the mathematical world took serious note of the ideas first put forth by Leibniz nearly 250 years earlier.

In this book we will seek to answer the question, "How can we become more logical?" We wish to apply logic not only to our formal course work but also to our everyday life. We need to be able to communicate intelligently with others; we need to be able to analyze the arguments of our lawmakers and leaders; we need to be intelligent consumers in analyzing the claims of advertisers. Whether we like it or not, logic is an important part of the world about us, and in this book we will lay the groundwork that will aid us in becoming more "logical."

1.2 Statements

Logic is a method of reasoning that accepts no conclusions except those that are inescapable. This is possible because of the strict way in which every concept is defined. That is, everything must be defined in a way that leaves no doubt or vagueness in meaning. Nothing can be taken for granted, and dictionary definitions are not usually sufficient. For example, in English a sentence may be defined as "a word or group of words stating, asking, commanding, requesting, or exclaiming something; conventional unit of connected speech or writing, usually containing a subject and predicate, beginning with a capital letter, and ending with an end mark."* In symbolic logic, however, a sentence has a much more restricted meaning and is called a statement.

STATEMENT

> A **statement** is a sentence that is either true or false but not both true and false.

If the sentence is a question or a command, or if it is too vague (or nonsensical), then it cannot be classified as true or false, so we would not call it a statement.

EXAMPLE 1

1. Neil Armstrong walked on the moon.
2. $3 + 2 = 7$.
3. Donald Duck is President.

All of these are statements, since they are either true or false. On the other hand, consider

4. Go away!
5. $3 + x = 7$.
6. What are you doing?

These are not statements by our definition, since they cannot satisfactorily be classified as either true or false. □

Difficulty in simplifying arguments may arise because of their length, the vagueness of the words they use, their literary style, or the possible emotional impact of the words they use. Consider the following two arguments.

1. If George Washington was assassinated, he is dead.
   Therefore, if he is dead, he was assassinated.
2. If you use heroin, you first used marijuana.
   Therefore, if you first used marijuana, you use heroin.

Logically, these two arguments are exactly the same, and both are *invalid* forms of reasoning. Nearly everyone would admit to the absurdity of the first, but many accept the second argument because of the emotional appeal of the words used.

---

* *Webster's New World Dictionary, College Edition* (World Publishing Company, 1960).

To avoid these difficulties and to aid us in simplifying complicated logical arguments, we can set up the artificial symbolic language we referred to earlier. The language we invent here is necessarily simpler than any natural language; it is a notational shorthand. We denote *simple statements with letters, such as p, q, r, s,* $\cdots$, and then we define certain connectives. Our goal, insofar as possible, is

1. **to translate the English statements into symbolic form**
2. **to simplify the symbolic form**
3. **to translate the simpler form back into English statements**

For the purposes of this text, we will assume that the translation back and forth between ordinary and symbolic language can be done in a straightforward manner. Actually, this not always the case, unfortunately. Ordinary language may have subtle relationships that go beyond the exact words of a single sentence. We should keep in mind the interpretations we give to our symbols in a particular problem. When translating, we must ask ourselves what the natural-language sentence means, and we must then try to find a symbolic-language statement that has, as nearly as possible, the same meaning.

We can avoid the problem of length by considering only simple statements connected by certain well-defined connectives, such as *not, and, or, neither . . . nor, if . . . then, unless, because,* and so on.

A *compound statement* is formed by combining simple statements with operators, or connectives. From our basic definition of a statement, we see that the *truth value* of any statement is either true (T) or false (F).

TRUTH VALUE

The **truth value** of a *simple statement* is either true (*T*) or false (*F*).

The **truth value** of a *compound statement* is true or false and depends only on the truth values of its simple compound parts. It is determined by using the rules for connecting those parts with well-defined operators, such as *and, or, not, if ... then, neither ... nor, unless,* and *because.*

We cannot assume that we know the meanings of the connectives *and, or, not,* and so on, even though they may seem obvious and simple. The strength of logic is that it does not leave any meanings to chance or to individual interpretation. In defining the truth values of these words, however, we will try to conform to common usage.

1.3 Conjunction

If p and q represent two simple statements, then the compound statement "p and q" uses the operator called **conjunction.** The word *and* is symbolized $\wedge$. For example, the statement

I have a penny in my pocket and I have a quarter in my pocket.

is a compound statement. When will this statement be true? Four actual possibilities exist:

1. I have a penny in my pocket. I have a quarter in my pocket.
2. I have a penny in my pocket. I do not have a quarter in my pocket.
3. I do not have a penny in my pocket. I have a quarter in my pocket.
4. I do not have a penny in my pocket. I do not have a quarter in my pocket.

Let p and q represent simple statements as follows:

p: I have a penny in my pocket.
q: I have a quarter in my pocket.

These four possibilities can be summarized as follows:

	p	q
1.	T	T
2.	T	F
3.	F	T
4.	F	F

We see that the only time the original statement

I have a penny in my pocket, and I have a quarter in my pocket.

is true is when p and q are both true; otherwise, it is false. Thus we define $p \wedge q$ according to Table 1.

Table 1
DEFINITION OF CONJUNCTION

p	q	$p \wedge q$
T	T	T
T	F	F
F	T	F
F	F	F

It is worth noting that statements p and q need not be related. For example,

Fish swim and Neil Armstrong walked on the moon.

is a true compound statement.

1.4 Disjunction

The operator *or,* denoted $\vee$, is called **disjunction.** The meaning of this word is ambiguous when used in everyday language, as we can see by considering the following examples. Let

p: Jack has a penny in his pocket.
q: Jack has a quarter in his pocket.
r: Jack is in Russia.
s: Jack is in Spain.

Jack now makes two statements:

I. I have a penny or a quarter in my pocket.
II. I am in Russia or in Spain.*

We will consider the intention of these statements.

In statement I, Jack seems to be claiming that at least one of the statements is true, but maybe both are true. On the other hand, the intention of statement II is that one and only one of the statements is true. He would not mean that he is in both Russia and Spain.

Statement I illustrates a use of the connective *or* that is called the *inclusive or* and is denoted by $\vee$. If p and q are two statements, we call $p \vee q$ *disjunction*; and in order to determine its truth or falsity, we must consider the truth or falsity of p and q. We therefore again consider the four possibilities to determine whether Jack was truthful in his assertion

I have a penny or a quarter in my pocket.

Case 1 (p is T; q is T): Jack has a penny in his pocket.
Jack has a quarter in his pocket.

Under these conditions we would not call Jack a liar, so we say that the compound statement is also true.

Case 2 (p is T; q is F): Jack has a penny in his pocket.
Jack does not have a quarter in his pocket.

Case 3 (p is F; q is T): Jack does not have a penny in his pocket.
Jack has a quarter in his pocket.

For cases 2 and 3 we would not call Jack a liar, so we say that these compound statements are also true.

Case 4 (p is F; q is F) Jack does not have a penny in his pocket.
Jack does not have a quarter in his pocket.

In this case we would call Jack a liar and say that his assertion is false.

Thus we define $p \vee q$ according to Table 2.

Table 2
DEFINITION OF DISJUNCTION

p	q	$p \vee q$
T	T	T
T	F	T
F	T	T
F	F	F

*We have used literary shorthand here. These sentences are to be interpreted as "I have a penny in my pocket, or I have a quarter in my pocket" and, "I am in Russia, or I am in Spain." We will use this type of shorthand throughout the text if no ambiguity results.

A disjunction is false only if both of its component statements are false.

Statement II illustrates a use of the connective *or* that is called the *exclusive or.* Let's now consider whether Jack is truthful when he makes the assertion

I am in Russia or in Spain.

Obviously it is not possible for Jack to be in both places at the same time. If Jack is in Russia, then he cannot be in Spain; if he is in Spain, he cannot be in Russia. Therefore the intent of Statement II is

I am either in Russia or in Spain.

and the statement is true if one and only one of the disjuncts is true. The truth table here is as follows:

r	s	either r or s
T	T	F
T	F	T
F	T	T
F	F	F

In this book we will use disjunction to mean *inclusive or.* If the intent of the statement requires *exclusive or,* we will translate the statement by using the connective *either . . . or.* **Remember**, *inclusive or* means *either or both,* while *exclusive or* means *either but* ***not*** *both.*

1.5 Negation

The denial of any proposition p is called *negation* and is symbolized by $\sim p$. Table 3 provides a straightforward definition of negation.

Table 3
DEFINITION OF NEGATION

p	$\sim p$
T	F
F	T

EXAMPLE 2

If t: Otto is telling the truth, then
$\sim t$: Otto is not telling the truth.

We can also translate $\sim t$ as "It is not the case that Otto is telling the truth." □

Any statement can be negated, but we need to be careful about the way we form the negation of a compound statement. The negation of

I have a penny and a quarter in my pocket.

is symbolized by $\sim (p \wedge q)$ and translated as

It is not the case that I have a penny and a quarter in my pocket.

The negation is not formed by negating each of the simple statements. In negating a statement, we must not change the statement itself. For example let

b: Jack's car is blue.
c: Jack's car is red.

Statement c is not the negation of statement b, even though both statements cannot be true. The negation of b is the statement

$\sim b$: Jack's car is not blue.

and the negation of c is the statement

$\sim c$: Jack's car is not red.

We must be careful when negating statements containing the words *all, none,* or *some.*

EXAMPLE 3 Write the negation of

All students have pencils.

Solution Did you think of "No students have pencils" or "All students do not have pencils"? Remember, if a statement is false, then its negation must be true. The correct negation is

Not all students have pencils.

or

At least one student does not have a pencil.

or

Some students do not have pencils. □

In mathematics, the word *some* is used to mean "at least one." Table 4 gives some of the common negations.

Table 4
NEGATION OF
ALL, SOME,* and *NO

Statement	Negation
All	Some ... not
Some	No
Some ... not	All
No	Some

EXAMPLE 4 Write the negation of each of the following statements.

a. All people have compassion.
b. Some animals are dirty.
c. Some students do not take Math 101.
d. No students are enthusiastic.

Solution

a. Some people do not have compassion.
b. No animal is dirty.
c. All students take Math 101.
d. Some students are enthusiastic. □

We will consider the negation of compound statements in the next chapter.

1.6 Translating and Combining Connectives

Working with logical arguments requires the ability to translate from English into symbols and from symbols back into English. For example the statement

Bonnie and Clyde are cute.

can be translated as follows. Let

b: Bonnie is cute.
c: Clyde is cute.

Then we have

$b \wedge c$: Bonnie is cute and Clyde is cute.

so $b \wedge c$ is the symbolic statement.

EXAMPLE 5 Translate the compound statement

Either Bonnie is cute or Clyde is cute.

into symbolic form.

Solution We see that the *either . . . or* is translated by the *exclusive or.* We can rewrite the sentence as

Bonnie is cute or Clyde is cute, but not both.

We translate the statement by

$$(b \vee c) \quad \wedge [\sim (b \wedge c)].$$

Notice that parentheses are used in this examplejust as they are in algebra: to indicate the order of operations. Thus, $\sim (b \wedge c)$ means the negation of the statement *b and c.* □

You do not only need to translate from English into logical symbols, but also from logical symbols back into English.

EXAMPLE 6 Suppose:

p: I eat spinach.
q: I am strong.

Then we wish to translate the following statements into words:

a.	$p \wedge q$	**b.**	$\sim p \vee q$
c.	$\sim p$	**d.**	$\sim (\sim q)$
e.	$\sim (p \vee q)$	**f.**	$p \wedge \sim q$

Solution Our translations would be as follows:

a. I eat spinach, and I am strong.
b. I do not eat spinach.
c. It is not the case that I eat spinach or that I am strong.
d. I do not eat spinach, or I am strong.
e. It is not the case that I am not strong [which means, of course, that I am strong].
f. I eat spinach, and I am not strong. □

1.7 PROBLEM SET 1

A

1. According to our definition, which of the following are statements?
   a. This banana is green.
   b. $3 + 7 = 10$.
   c. The difference between two natural numbers is a natural number.
   d. If I spend $1,000 a day for 2,000 years, I will not have spent a billion dollars.
   e. Are you happy?

2. According to our definition, which of the following are statements?
   a. Tomorrow the world will come to an end!
   b. Tom must pay his debts unless he wants to go to jail.
   c. Thomas Jefferson was the twenty–third president of the United States.
   d. Sit down and be quiet.
   e. Shakespeare was the greatest writer the world has ever known.

3. According to our definition, which of the following are statements?
   a. Division by zero is impossible.
   b. If Otto does not like to do long division, then he is perfectly normal.
   c. If Farley were to discover a formula for finding prime numbers, he would be famous.
   d. Abstraction exchanges fuzzy, yellow.
   e. Logic is not as difficult as I had anticipated.

4. According to our definition, which of the following are statements?
   a. Hickory dickory dock, the mouse ran up the clock.
   b. Is John ugly?
   c. John has a wart on the end of his nose.
   d. Dan and Mary were married on August 3, 1989.
   e. Do you have a cold?

5. Define conjunction.

6. Define disjunction.

7. Define negation.

Write the negation of each statement in Problems 8–10.

8. **a.** All mathematicians are ogres.
   b. Some integers are negative.
   c. Some people do not pay taxes.
   d. No even integers are divisible by 5.

9. **a.** All counting numbers are divisible by 1.
   b. Some apples are rotten.
   c. Some integers are not odd.
   d. No triangles are squares.

10. **a.** All women are intelligent.
   b. Some rectangles are squares.
   c. Some rectangles are not squares.
   d. No nice people are dangerous.

Let p represent the statement "Bonnie is cute" and q represent "Clyde is cute." Translate each of the statements in Problems 11–15 into symbols. (Assume that ugly and cute are opposites; that is, to be "not ugly" is to be cute.)

EXAMPLE 7 Both Bonnie and Clyde are ugly.

Solution

$\sim p$: Bonnie is ugly.
$\sim q$: Clyde is ugly.
Thus $(\sim p) \wedge (\sim q)$. □

11. Both Bonnie and Clyde are cute.
12. It is not the case that both Bonnie and Clyde are cute.
13. Bonnie is cute, or Clyde is ugly.
14. Bonnie is cute, and Clyde is ugly.
15. Either Bonnie is cute or Clyde is cute, but not both.

Translate each of the statements in Problems 16–20 into words. Let

p: Paul is peculiar.
q: Paul likes to read math textbooks.

EXAMPLE 8 $(\sim p) \wedge q$

Solution

$\sim p$: Paul is not peculiar. Thus the statement is "Paul is not peculiar, and he likes to read math textbooks." □

16. $p \wedge q$
17. $\sim p \wedge q$
18. $\sim (p \wedge q)$
19. $p \vee (\sim q)$
20. $\sim p \vee (\sim q)$

Translate the statements in Problems 21–25 into symbols. For each simple statement, be sure to indicate the meanings of the symbols you use.

21. Pat is eating, drinking, and having a good time.
22. Sam will not accept the nomination.
23. You will not go tonight, and you will not go tomorrow.
24. Fat Albert lives to eat, and he does not eat to live.
25. The decision will depend on judgment or intuition, not on who paid the most.

Write T (true) or F (false) for each statement in Problems 26–31. Modify each false statement to make it true.

26. For all real numbers x, $|x| > 0$.

27. For all real numbers x, $\sqrt{x^2} = x$.

28. For some real number x, $x^2 \leq 0$.

29. For some real number x, $(x + 3)^2 = x^2 + 6x + 9$.

30. There is no real number x, $|x + 4| \leq 0$.

31. There is no real number x, $x^2 < 0$.

CHAPTER 2 TRUTH TABLES

2.1 Definition and Example

Much of what has been said about logic and the relationships between propositions of reasoning can be illustrated by means of ***truth tables.*** A truth tables is a device that shows how the truth values of compound statements depend on the connectives used and on the truth values of the component simple statements. Tables 1, 2, and 3 in Chapter 1 are fundamental for constructing truth tables. We summarize them in Tables 5 and 6.

Table 5
FUNDAMENTAL CONNECTIVES

Connective	Symbol	Name	Symbolic Statement	Example
and	$\wedge$	conjunction	$p \wedge q$	Fish swim *and* birds fly.
or	$\vee$	disjunction	$p \vee q$	Fish swim *or* birds fly.
not	$\sim$	negation	$\sim p$	Fish do *not* swim.

Table 6
TRUTH TABLE FOR THE FUNDAMENTAL CONNECTIVES

p	q	$p \wedge q$	$p \vee q$	$\sim p$	$\sim q$
T	T	T	T	F	F
T	F	F	T	F	T
F	T	F	T	T	F
F	F	F	F	T	T

EXAMPLE 1

Construct a truth table for the compound statement:

Alfie did not come last night, and he did not pick up his money.

Solution

Let p: Alfie came last night.
q: Alfie picked up his money.

Then the statement can be written $(\sim p) \wedge (\sim q)$. To begin, we list all the possible combinations of truth values for the simple statements p and q.

p	q
T	T
T	F
F	T
F	F

Then we insert the truth values for $\sim p$ and $\sim q$.

p	q	$\sim p$	$\sim q$
T	T	**F**	**F**
T	F	**F**	**T**
F	T	**T**	**F**
F	F	**T**	**T**

Finally, we insert the truth values for $(\sim p) \wedge (\sim q)$. To do this, we label a column "$(\sim p) \wedge (\sim q)$" at the top and then compare the columns $\sim p$ and $\sim q$ according to the definition of conjunction to find the entries of $(\sim p) \wedge (\sim q)$ as shown by the arrows.

conjunction
↓

p	q	$\sim p$	$\sim q$	$(\sim p) \wedge (\sim q)$
T	T	F	F	**F**
T	F	F	T	**F**
F	T	T	F	**F**
F	F	T	T	**T**
		↓	↓	↑

Compare these entries with
the definition of *conjunction*
to obtain the entries in the last column. □

EXAMPLE 2 Construct a truth table for $\sim (\sim p)$.

Solution We begin with p.

p
T
F

Next we fill in the values for $\sim p$ according to the definition of negation.

p	$\sim p$
T	**F**
F	**T**

Finally, we fill in the values for $\sim (\sim p)$. The arrows show how we obtain the entries for this column.

negation
↓

p	$\sim p$	$\sim (\sim p)$
T	F	**T**
F	T	**F**
	↑	

Look at this column. Use the definition of *negation* for all the truth values in this column. □

Notice that $\sim (\sim p)$ and p have the same truth values. If two propositions have the same truth values, one can replace the other in any logical expression. This means that the double negative of a proposition is the same as the original proposition.

LAW OF DOUBLE NEGATION

> *$\sim(\sim p)$ may be replaced by p in any logical expression.*

EXAMPLE 3 Construct a truth table to determine when the following statement is true.

$$\sim(p \wedge q) \wedge [(p \vee q) \wedge q]$$

Solution We begin as before and move from left to right, focusing our attention on only two columns at a time (refer to Table 5 to find the correct entries).

Step 1. Attend to the parentheses first. Look at Columns A and B below, along with the definition of conjunction, to fill in the entries of Column C.

Step 2. Use Column C and the definition of negation to fill in the entries of Column D.

Step 3. Use Columns A and B and the definition of disjunction to fill in the entries of Column E.

Step 4. Use Column E and B and the definition of conjunction to fill in the entries of Column F.

Step 5. Use Columns D and F, which are the left and right sides of the final operation, and the definition of conjunction to obtain the entries of Column G.

A	B	C	D	E	F	G
p	q	$p \wedge q$	$\sim(p \wedge q)$	$p \vee q$	$(p \vee q) \wedge q$	$\sim(p \wedge q) \wedge [(p \vee q) \wedge q]$
T	T	T	F	T	T	F
T	F	F	T	T	F	F
F	T	F	T	T	T	T
F	F	F	T	F	F	F

□

2.2 PROBLEM SET 2

A

Find the truth table for the statements given in Problems 1–10.

1. $\sim p \wedge (\sim q)$ **2.** $\sim(p \wedge q)$
3. $\sim p \vee (\sim q)$ **4.** $p \wedge (p \vee q)$
5. $r \wedge (\sim s)$ **6.** $s \vee (\sim s)$
7. $(r \wedge s) \vee (\sim s)$ **8.** $(\sim r) \vee (\sim s)$
9. $(p \vee q) \vee (p \wedge q)$ **10.** $(\sim p \wedge q) \vee (\sim q)$
11. $[p \wedge (\sim q)] \wedge p$
12. $\{p \wedge [q \vee (\sim p)]\} \vee [p \wedge (\sim q)]$
13. $(p \wedge q) \wedge r$
14. $(p \wedge q) \wedge (\sim r)$
15. $[(p \vee q) \wedge (\sim r)] \wedge r$
16. $\{p \wedge [q \vee (\sim p)]\} \vee r$

17. Let p: $2 + 3 = 5$.
q: The moon is made of green cheese.
Tell which of the statements in Problems 1–4 are true.

18. Let r: Mankind pollutes the environment.
s: Mankind will survive.
Translate each of the statements in Problems 5–8 into words, and tell under which conditions each will be true.

19. If r is T and s is F, is the statement in Problem 7 true?

20. If r is F and s is T, is the statement in Problem 8 true?

21. If p is T and q is T, is the statement in Problem 9 true?

22. If p is F and q is F, is the statement in Problem 10 true?

23. **a.** How many entries are there on a truth table for one proposition?
b. How many entries are there on a truth table for two propositions?
c. How many entries are there on a truth table for three propositions (see Problem 10)?

24. How many entries are there on a truth table for four propositions?

25. Consider the patterns found in Problems 23 and 24. How many entries are there on a truth table for n propositions?

CHAPTER 3 THE CONDITIONAL

3.1 If–Then

The statement "if p, then q" is called a **conditional**. It is symbolized by $p \rightarrow q$; p is called the *hypothesis* or **antecedent**, and q is called the *conclusion* or **consequent**. The situation for the conditional is similar to the situation we encountered for the word *or*. That is, in English we use the if–then connective in several ways. Consider the following example illustrating different senses of the conditional.

1. We can use if–then to indicate a *logical* relationship — one in which the consequent follows logically from the antecedent.

   If $\sim(\sim p)$ *has the same truth value as* p*, then* p *can replace* $\sim(\sim p)$*.*

2. We can use if–then to indicate a causal relationship.

   If John drops that rock, then it will land on my foot.

3. We can use if–then to report a decision on the part of the speaker.

   If John drops that rock, then I will hit him.

4. We can use if–then when the consequent follows from the antecedent by the very definition of the words used.

   If John drives an Oldsmobile, then John drives a car.

5. Finally, we can use if–then to make a *material implication.* Sometimes the conditional is used when there is no logical, causal, or definitional relationship between the antecedent and consequent. It is often used in a humorous or emphatic way:

   If John got an A on that test, then I'm a monkey's uncle.

   The consequent is obviously false, and the speaker wishes to emphasize that the antecedent is also false.

Our task is to try to devise a definition of the conditional that will apply in all these different types of if–then statements. We will approach the problem by asking under what circumstances a given conditional would be false. Let's consider another example.

Suppose I make you a promise. "If I win the contest, then I will give you $10." If I keep my promise, we say the statement is true; if I break my promise, we say it is false. Let

p: I win the contest.
q: I will give you \$10.

We symbolize the promise by $p \rightarrow q$. There are four possibilities:

	p	q	
Case 1:	T	T	I win the contest; I will give you \$10.
Case 2.	T	F	I win the contest; I will not give you \$10.
Case 3:	F	T	I do not win the contest; I will give you \$10.
Case 4:	F	F	I do not win the contest; I will not give you \$10.

When will I have broken my promise?

The only time you can say that I will have broken my promise is in case 2. The test for a conditional is to determine when it is false. In symbols,

$p \rightarrow q$ is false whenever $p \wedge (\sim q)$ is true *This is from case 2.*

or

$p \rightarrow q$ is true whenever $p \wedge (\sim q)$ is false

This is the same as saying

$p \rightarrow q$ is true whenever $\sim [p \wedge (\sim q)]$ is true

Let's construct a truth table for $\sim [p \wedge (\sim q)]$.

p	q	$\sim q$	$p \wedge (\sim q)$	$\sim [p \wedge (\sim q)]$
T	T	F	F	T
T	F	T	T	F
F	T	F	F	T
F	F	T	F	T

We use this truth table to define the conditional $p \rightarrow q$, as shown in Table 7.

Table 7
DEFINITION OF THE CONDITIONAL

p	q	$p \rightarrow q$
T	T	T
T	F	F
F	T	T
F	F	T

The following examples illustrate the definition of the conditional, as shown in Table 7.

EXAMPLE 1

Case 1: T → T

If $7 < 14$, *then* $7 + 2 < 14 + 2$.

This is a true statement, since both component parts are true. □

EXAMPLE 2

Case 2: T → F

If $7 + 5 = 12$, *then* $7 + 10 = 15$.

This is a false statement, since the antecedent is true but the consequent is false. □

EXAMPLE 3

Case 3: F $\rightarrow$ T

If you have six legs, then George Washington was president.

True, since the consequent is true (and if the consequent is true, the whole implication is true, whether or not the antecedent is true). □

Example 3 shows that the conditional, in mathematics, does not imply any cause-and-effect relationship. *Any* two statements can be connected with the conditional, and the result must be T or F.

EXAMPLE 4

Case 4: F $\rightarrow$ F

If $16 = 8$, *then* $8 = 4$.

This is a true statement, since both component parts are false. □

The *if*-part of an implication need not be stated first. *All* of the following statements have the *same meaning*:

1. If I win the contest, then I will give you $10.
2. Winning the contest implies that I will give you $10.
3. I will give you $10 if I win the contest.
4. I win the contest only if I will give you $10.
5. For me to give you $10, it is sufficient that I win the contest.
6. Winning the contest makes it necessary for me to give you $10.
7. I will give you $10 provided that I win the contest.
8. I will give you $10 when I win the contest.
9. I will give you $10 whenever I win the contest.

We can state these nine equivalent statements in symbolic form as follows:

1. if p, then q
2. $p \rightarrow q$
3. q if p
4. p only if q
5. p is sufficient for q
6. q is necessary for p
7. q provided p
8. q when p
9. q whenever p

Some statements, although not originally written as a conditional, can be put into if–then form. For example, "All ducks are birds" can be rewritten as, "If it is a duck, then it is a bird." Thus we add one additional form to the list:

10. all d are b *Translation: All ducks are birds.*

EXAMPLE 5

Translate the following sentence, taken from a tax form, into symbolic form.

If you do not itemize deductions on Schedule A and you have charitable contributions, then complete the worksheet on page 14 and enter the allowable part on line 34b.

Solution

Step 1. Isolate the simple statements and assign them variables. Let

d: You itemize deductions on Schedule A.
c: You have charitable contributions.
w: You complete the worksheet on page 14.
b: You enter the allowable part on line 34b.

Step 2. Rewrite the sentence, making substitutions for the variables.

If $\sim d \wedge c$, then $(w \wedge b)$

Step 3. Complete the translation into symbols.

$$(\sim d \wedge c) \rightarrow (w \wedge b)$$ □

The if–then can take on several meanings, but the $\rightarrow$ has only one meaning — the one we defined in Table 7. What $p \rightarrow q$ means is $\sim [p \wedge (\sim q)]$. However, in practice we use $p \rightarrow q$ to represent if–then or implies statements from English. How can we justify this? As we have seen, part of the difficulty in translating from English into symbols lies in stripping away the ambiguity of the English language. However, the validity of all valid arguments of the if–then type used in this text is *preserved,* even if the additional meanings of the conditional are ignored. We will discuss this point more fully later, when we consider validity and valid arguments. For the present we will translate all the various if–then forms according to the $p \rightarrow q$ defined in Table 7.

3.2 Converse, Inverse, and Contrapositive

Related to the implication $p \rightarrow q$ are other statements, which we will now define.

VARIATIONS OF THE CONDITIONAL

Given the conditional $p \rightarrow q$, we define:

1. The *converse:* $q \rightarrow p$
2. The *inverse:* $\sim p \rightarrow \sim q$
3. The *contrapositive:* $\sim q \rightarrow \sim p$

EXAMPLE 6

Let p: This animal is a bird.
q: This animal has wings.

Given $p \rightarrow q$, write its converse, inverse, and contrapositive.

Solution

STATEMENT:	If this animal is a bird, then it has wings.
CONVERSE:	If this animal has wings, then it is a bird.
INVERSE:	If this animal is not a bird, then it does not have wings.
CONTRAPOSITIVE:	If this animal does not have wings, then it is not a bird. □

The original statement need not take the form $p \to q$, and the antecedent or the consequent can be any statement. When we write the converse, inverse, or contrapositive, a double negative may result. In this case the double negative should be replaced by the original statement, using the law of double negation.

EXAMPLE 7 Given $p \to \sim q$, write its converse, inverse, and contrapositive.

Solution

CONVERSE: $\sim q \to p$
INVERSE: $\sim p \to q$
CONTRAPOSITIVE: $q \to (\sim p)$ □

EXAMPLE 8 Given $\sim p \to \sim q$, write its converse, inverse, and contrapositive.

Solution

CONVERSE: $\sim q \to (\sim p)$
INVERSE: $p \to q$
CONTRAPOSITIVE: $q \to p$ □

Not all these statements are equivalent in meaning, as can be seen by considering Example 9.

EXAMPLE 9 Write the converse, inverse, and contrapositive of the statement: If it is an Oldsmobile, than it is a car.

Solution

Let p: It is an Oldsmobile.
q: It is a car.

Then the given statement is written $p \to q$.

CONVERSE: $q \to p$: If it is a car, then it is an Oldsmobile.

INVERSE: $\sim p \to (\sim q)$: If it is not an Oldsmobile, then it is not a car.

CONTRAPOSITIVE: $\sim q \to (\sim p)$: If it is not a car, then it is not an Oldsmobile. This is a true statement, *provided that* the original statement is true. □

Example 9 shows that the converse of a true statement need not be true. It may be, but it need not be. Notice from Table 8 that the contrapositive and the direct statement always have the same truth value, as do the converse and inverse. Hence an implication may always be replaced by its contrapositive, without affecting its truth or falsity.

Table 8
TRUTH TABLE FOR THE CONVERSE, INVERSE, AND CONTRAPOSITIVE OF A GIVEN STATEMENT

p	q	$\sim p$	$\sim q$	Statement $p \to q$	Converse $q \to p$	Inverse $\sim p \to (\sim q)$	Contrapositive $\sim q \to (\sim p)$
T	T	F	F	T	T	T	T
T	F	F	T	F	T	T	F
F	T	T	F	T	F	F	T
F	F	T	T	T	T	T	T

LAW OF CONTRAPOSITION

A statement may be replaced by its contrapositive without having its truth value affected.

EXAMPLE 10

Let p: John obeys the law.
q: John goes to jail.

STATEMENT: If John obeys the law, then he does not go to jail.

The statement $p \rightarrow (\sim q)$ can be replaced by its contrapositive, $q \rightarrow (\sim p)$, in any logical expression.

CONTRAPOSITIVE: If John goes to jail, then he does not obey the law.

The statement *cannot* be replaced by the converse or the inverse.

CONVERSE: If John does not go to jail, then he obeys the law.

INVERSE: If John does not obey the law, then he goes to jail. □

3.3 PROBLEM SET 3

A

Translate the sentences in Problems 1–14 into if–then form (in English), and then translate them into symbolic form. Be sure to indicate the letters you use for each simple statement.

1. You will succeed only if you value others' judgments.
2. I will go Saturday if I get paid.
3. Everything happens to everybody sooner or later if there is time enough. (G. B. Shaw)
4. All work is noble. (Thomas Carlyle)
5. $|f(x) - L| < \epsilon$ whenever $|x - a| < \delta$.
6. The presence of oxygen is necessary for combustion.
7. Combustion is a sufficient condition for the presence of oxygen.
8. John is reasonable only if he knows all the facts.
9. We are not weak if we make a proper use of those means which the God of Nature has placed in our power. (Patrick Henry)
10. $x < 5$ provided x is non-negative.
11. $y^2 = 4$ is a necessary condition for $y = 2$.
12. For a polygon to be a square, it is necessary for it to be a rectangle.
13. All triangles are polygons.
14. All primes greater than 2 are odd primes.

By looking at the truth of the simple statements, tell whether each of the compound statements in Problems 15–19 is true.

15. If $7 + 9 = 16$, then $16 - 9 = 7$.
16. If $(-2)(-3) = -6$, then $(2)(3) = -6$.
17. You are a millionaire only if Mickey Mouse is under 65.
18. $15 + 3 = 18$, if $72 - (-10) = 82$.
19. $3 \cdot 5 = 15$, if $58 - (-8) = 50$.

In Problems 20–24, let

p: You like mathematics.

q: You like this text.

Translate each of the following into words.

20. $p \rightarrow q$
21. $\sim p \vee q$
22. $\sim q \rightarrow (\sim p)$
23. $[p \vee (\sim p)] \rightarrow q$
24. $\sim p \rightarrow q$

Write the truth table for each statement in Problems 25–37.

25. $p \rightarrow (\sim p \rightarrow q)$
26. $(p \vee q) \vee [p \wedge (\sim q)]$
27. $(p \wedge q) \wedge [p \rightarrow (\sim q)]$
28. $(p \wedge q) \rightarrow p$
29. $(p \rightarrow p) \rightarrow [q \rightarrow (\sim q)]$
30. $\sim p \rightarrow [\sim (p \wedge q)]$
31. $(p \rightarrow q) \rightarrow [\sim q \rightarrow (\sim p)]$
32. $(p \wedge q) \rightarrow (p \vee q)$
33. $(p \rightarrow q) \rightarrow [(p \vee q) \rightarrow q]$
34. $[(p \rightarrow q) \rightarrow p] \rightarrow (p \rightarrow q)$
35. $[(p \rightarrow q) \rightarrow r] \rightarrow [(q \rightarrow p) \rightarrow r]$
36. $[p \vee (\sim p \wedge q)] \rightarrow (p \vee q)$
37. $(p \vee q) \rightarrow [p \vee (\sim p \wedge q)]$

Write the converse, inverse, and contrapositive of each of the statements in Problems 38–46.

38. If you break the law, then you go to jail.
39. If p, then q.
40. $\sim p \rightarrow (\sim q)$
41. The sun shines only if you are happy.
42. If your car is not air-conditioned, you do not have friends.
43. I will go on Saturday, if I get paid.
44. If wishes were horses, then beggars would ride.
45. If you brush your teeth with Smiles toothpaste, then you have fewer cavities.
46. Your grandchildren will not survive if we do not stop the pollution of our environment.

The following statement was found on the residency questionnaire for the California State University system:

"Complete and submit this residency questionnaire only if your response to item 34a of part A was other than 'California' or your response to item 37 was 'no'."

Problems 47–51 refer to this statement.

47. Write the statement in symbolic form.
48. Are the instructions on the questionnaire correctly stated?
49. Let us rephrase the given statement as "If your response to item 34a of part A was other than 'California' or your response to item 37 was 'no,' then complete and submit this residency questionnaire." Translate this statement into symbolic form.

50. This is a continuation of Problem 49. The following two questions determine whether the residency questionnaire should be completed.

		Answer	
Question 34a:	Were you born in California?	Yes	No
Question 37:		Yes	No

There are only four possible responses to the two questions; list them. Using the symbolic statement for Problem 49 complete a truth table where p is true if Question 34a is "yes" and false if it is "no," and q is true if Question 37 is "yes" and false if it is "no."

51. This is a continuation of Problems 49–50. Rewrite (in English) the statement on the residency questionnaire so that it is easier to decide whether to fill out the form.

CHAPTER 4

ADDITIONAL OPERATORS

4.1 Biconditional

In Chapter 3 we took great care to point out that a statement $p \rightarrow q$ and its converse $q \rightarrow p$ do not have the same truth values. However, it may be the case that $p \rightarrow q$ *and also* $q \rightarrow p$. In this case we write

$$p \leftrightarrow q$$

and call this connective the **biconditional.** To determine the truth values of the biconditional, we must construct a truth table for $(p \rightarrow q) \wedge (q \rightarrow p)$.

p	q	$p \rightarrow q$	$q \rightarrow p$	$(p \rightarrow q) \wedge (q \rightarrow p)$
T	T	T	T	T
T	F	F	T	F
F	T	T	F	F
F	F	T	T	T

This leads us to define the biconditional so that it is true only when both p and q are true or when both p and q are false (that is, whenever they have the same truth values). This is shown in Table 9.

Table 9
DEFINITION OF THE BICONDITIONAL

p	q	$p \leftrightarrow q$
T	T	T
T	F	F
F	T	F
F	F	T

In mathematics, $p \leftrightarrow q$ is translated in several ways, all of which have the same meaning.

TRANSLATIONS FOR THE BICONDITIONAL

1. p if and only if q
2. q if and only if p
3. If p, then q, and conversely
4. If q, then p, and conversely
5. p is a necessary and sufficient condition for q
6. q is a necessary and sufficient condition for p

EXAMPLE 1

Rewrite the following into one statement:

1. If a polygon has three sides, then it is a triangle.
2. If a polygon is a triangle, then it has three sides.

Solution

A polygon is a triangle *if and only if* it has three sides. □

There is one additional way to denote the biconditional. If two propositions have the same truth values, we say that they are **equivalent**. Notice that $p \leftrightarrow q$ is true whenever both p and q have the same truth values. Thus we add another way of writing the biconditional to our list:

EQUIVALENCE

If p and q have the same truth values, we say that **p is equivalent to q**, and we write $p \leftrightarrow q$.

EXAMPLE 2

Show that $p \rightarrow q$ is equivalent to $\sim p \vee q$.

Solution

To establish equivalence, we must show that $(p \rightarrow q) \leftrightarrow (\sim p \vee q)$ is always true.

p	q	$p \rightarrow q$	$\sim p$	$\sim p \vee q$	$(p \rightarrow q) \leftrightarrow (\sim p \vee q)$
T	T	T	F	T	T
T	F	F	F	F	T
F	T	T	T	T	T
F	F	T	T	T	T

Since the result shows that we get all Ts, we say that the two statements are equivalent. □

4.2 Miscellaneous Operators

Occasionally we encounter other connectives, which also require precise definitions. We define them in terms of previous connectives, as shown in Table 10.

Table 10
ADDITIONAL OPERATORS

Connective	Definition	Comments
Either p or q	$(p \vee q) \wedge [\sim (p \wedge q)]$	This is the *exclusive or* discussed earlier, which says that p occurs or q occurs, but *not both.*
Neither p nor q	$\sim p \wedge (\sim q)$	Neither p nor q.
p unless q	$\sim q \rightarrow p$	This says p unless q. If q does not occur, then p will occur.
p because q	$q \rightarrow p$	This is the converse of the statement $p \rightarrow q$. It means that q implies p.
No p is q	$p \rightarrow (\sim q)$	This says that, if it is a p, then it cannot be a q.

The truth table for each of these new connectives is shown in Table 11.

Table 11
TRUTH TABLE FOR ADDITIONAL OPERATORS

p	q	**Either p or q** $(p \vee q) \wedge [\sim(p \wedge q)]$	**Neither p nor q** $\sim p \wedge (\sim q)$	**p unless q** $\sim q \rightarrow p$	**p because q** $q \rightarrow p$	**No p is q** $p \rightarrow (\sim q)$
T	T	F	F	T	T	F
T	F	T	F	T	T	T
F	T	T	F	T	F	T
F	F	F	T	F	T	T

4.3 PROBLEM SET 4

In Problems 1-6 write each of the pairs of statements as one statement, by using the appropriate connective.

EXAMPLE 3

i. If Mr. C. D. Money can buy a new car, then he can afford it.
ii. If Mr. C. D. Money can afford a new car, then he can buy it.

Solution

Mr. C. D. Money can buy a new car if and only if he can afford it. □

1. i. If $\triangle ABC$ is a right triangle, then $a^2 + b^2 = c^2$.
ii. If $a^2 + b^2 = c^2$, then $\triangle ABC$ is a right triangle.

2. i. If the prosecution can prove Otto's guilt, then the judge sentences him.
ii. If the judge sentences Otto, then the prosecution can prove his guilt.

3. i. If $x \cdot y = 0$, then $x = 0$ or $y = 0$.
ii. If $x = 0$ or $y = 0$, then $x \cdot y = 0$.

4. i. If P is prime, then it is a natural number with exactly two divisors.
ii. If P is a natural number with exactly two divisors, then it is prime.

5. i. If the United States develops greater strategic weapons, then Russia will do the same.
ii. If Russia develops greater strategic weapons, then the United States will do the same.

6. i. If a triangle is equilateral, then it is equiangular.
ii. If a triangle is equiangular, then it is equilateral.

Write each of the sentences in Problems 7–11 in symbolic form, where p: This problem can be solved; *and* q: Ron likes this problem.

7. Ron likes this problem, unless this problem cannot be solved.

8. Ron does not like this problem, because this problem cannot be solved.

9. It is not the case that this problem can be solved, and Ron likes this problem.

10. This problem cannot be solved, and Ron likes this problem.

11. If this problem can be solved, then Ron likes this problem, and this problem can be solved; therefore Ron likes this problem.

Translate the symbolic statements in Problems 12–16 into words. Let p and q be defined as in Problems 7–11.

12. $p \rightarrow q$
13. $q \rightarrow p$
14. $\sim q \wedge (\sim p)$
15. $(p \vee q) \wedge [\sim (p \wedge q)]$
16. $[(p \rightarrow q) \wedge (\sim q)] \rightarrow (\sim p)$

17. Show, by using truth tables, that $(p \rightarrow q)$ is equivalent to $(\sim p \vee q)$.
18. Show, by using truth tables, that $[p \rightarrow (\sim q)]$ is equivalent to $[\sim p \vee (\sim q)]$.

Find the truth value of each statement in Problems 19–21.

19. $6 \cdot 9 = 15$ if and only if man walked on the moon.
20. $7 \cdot 5 = 35$ if and only if water runs uphill.
21. $\sim p \leftrightarrow (p \vee q)$ when p is T and q is F.

22. Verify the results of Table 11 for the connective *unless* by using the definition given in Table 10.
23. Verify the results of Table 11 for the connective *neither . . . nor* by using the definition given in Table 10.
24. Verify the results of Table 11 for the connective *either . . . or* by using the definition given in Table 10.
25. A rather obscure operator called Sheffer's Stroke is written "$p|q$" and is defined by $\sim p \vee (\sim q)$. Construct a truth table for $p|q$.
26. All the other connectives can be defined in terms of Sheffer's Stroke (see Problem 25). Write *and, or,* and *not* in terms of Sheffer's Stroke.

Translate the statements in Problems 23–37 into symbolic form.

27. $x = 10$ iff $2x = 20$.
28. $3x + 2 = 11$ is equivalent to $x = 3$.
29. Neither smoking nor drinking is good for your health.
30. I will not buy a new house unless all provisions of the sale are clearly understood.
31. No man is an island.
32. A necessary and sufficient condition for a function to be differentiable is that it is continuous.
33. I cannot go with you, because I have a previous engagement.
34. Either I will invest my money in stock or I will put it in a savings account.
35. If a triangle has two sides equal, then it is isosceles, and conversely.
36. No man who has once heartily and wholly laughed can be altogether irreclaimably bad. (Thomas Carlyle)
37. In a moral point of view, revolution is justified if by the mere force of numbers a majority should deprive a minority of any clearly written constitutional right. (Abraham Lincoln)

38. Indicate which of the following are equivalent to the statement

If Herbert wins the contest, then he cheated.

(Assume that "Herbert is honest" is the same as "Herbert is not a cheater.")

a. Herbert wins the contest and is a cheater.
b. Herbert, an honest person, did not win the contest.
c. If Herbert is honest then he did not win the contest.
d. It is not the case that Herbert wins or is a cheater.
e. If Herbert is a cheater, then he wins the contest.

39. Indicate which of the following are equivalent to the statement

If Robin is guilty, then she was at the scene of the crime.

(Assume that "Robin is not guilty" is the same as "Robin is innocent.")
a. If Robin was at the scene of the crime, then she is innocent.
b. Robin is innocent or was at the scene of the crime.
c. If Robin was not at the scene of the crime, then she is innocent.
d. If Robin was at the scene of the crime, then she is guilty.
e. Robin is guilty and was not at the scene of the crime.
f. If Robin is innocent, then she was not at the scene of the crime.

CHAPTER 5 TAUTOLOGIES

5.1 Definition

Many statements in symbolic logic are *always true.* For example, the compound statement

$$p \vee (\sim p)$$

is always true, regardless of the truth value of p. Statements that are always true are called tautologies.

TAUTOLOGY

A **tautology** is a statement that is true for every case of its truth table.

To prove that a given statement is a tautology, we construct a truth table and make sure that the final column (representing the given statement) consists of all Ts. If there is even one F, it is not a tautology. If the values in the last column are all Fs, the given statement is called a **contradiction.**

EXAMPLE 1 Show that $(p \wedge q) \vee [p \rightarrow (\sim q)]$ is a tautology.

Solution We construct the truth table.

p	q	$p \wedge q$	$\sim q$	$p \rightarrow \sim q$	$(p \wedge q) \vee [p \rightarrow (\sim q)]$
T	T	T	F	F	T
T	F	F	T	T	T
F	T	F	F	T	T
F	F	F	T	T	T

Thus, since the statement is true for every case of its truth table, it is a tautology. □

EXAMPLE 2 Is $(p \vee q) \rightarrow (\sim q \rightarrow p)$ a tautology?

Solution Construct the truth table.

p	q	$p \vee q$	$\sim q$	$\sim q \rightarrow p$	$(p \vee q) \rightarrow (\sim q \rightarrow p)$
T	T	T	F	T	T
T	F	T	T	T	T
F	T	T	F	T	T
F	F	F	T	F	T

Thus it is a tautology. □

5.2 Implication

If a conditional is a tautology, as in Example 2, then it is called an **implication** and is symbolized by $\Rightarrow$. That is, Example 2 can be written

$$(p \vee q) \Rightarrow (\sim q \rightarrow p)$$

The implication symbol $p \Rightarrow q$ is pronounced "p implies q."

A biconditional statement $p \leftrightarrow q$ that is also a tautology is a **logical equivalence**, written $p \Leftrightarrow q$ and read "p is logically equivalent to q."

EXAMPLE 3 Show that $(p \rightarrow q) \Leftrightarrow [\sim p \wedge (\sim q)]$.

Solution We construct the truth table.

p	q	$p \rightarrow q$	$\sim q$	$p \wedge (\sim q)$	$\sim [p \wedge (\sim q)]$	$(p \rightarrow q) \leftrightarrow [\sim p \wedge (\sim q)]$
T	T	T	F	F	T	T
T	F	F	T	T	F	T
F	T	T	F	F	T	T
F	F	T	T	F	T	T

Since all possibilities are true, it is a logical equivalence and we write

$$(p \rightarrow q) \Leftrightarrow [\sim p \wedge (\sim q)]$$ □

Notice the difference between the symbols $\rightarrow$ and $\Rightarrow$, as well as between $\leftrightarrow$ and $\Leftrightarrow$. The conditional ($\rightarrow$) and biconditional ($\leftrightarrow$) are used as logical connectives, and the truth table for these may be true or false. If, because of the statements being connected, the truth tables give *only true values*, then we use the symbols $\Rightarrow$ and $\Leftrightarrow$ in place of $\rightarrow$ and $\leftrightarrow$ and call them implication ($\Rightarrow$) or logical equivalence ($\Leftrightarrow$).

5.3 Laws of Logic

Tautologies are very important in logic and mathematics. In the remainder of this section we will consider eight tautologies, or laws of logic; you should become familiar with these so that you can refer to each by name. They will form the foundation upon which we can build certain methods of proof that are important in logic and mathematics. We will first list the eight tautologies and then discuss and prove each.

TAUTOLOGY	SYMBOLIC FORM
1. Trivial tautology	1. $p \Leftrightarrow p$
2. Law of double negation	2. $p \Leftrightarrow \sim(\sim p)$
3. Law of the excluded middle	3. $p \vee \sim q$
4. Direct reasoning	4. $[(p \rightarrow q) \wedge p] \Rightarrow q$
5. Indirect reasoning	5. $[(p \rightarrow q) \wedge \sim q] \Rightarrow \sim p$
6. Law of transitivity	6. $[(p \rightarrow q) \wedge (q \rightarrow r)] \Rightarrow (p \rightarrow r)$
7. Law of contraposition	7. $(p \rightarrow q) \Leftrightarrow (\sim q \rightarrow \sim p)$
8. Disjunctive syllogism	8. $[(p \vee q) \wedge \sim p] \Rightarrow q$

5.4 The Trivial Tautology, $p \Leftrightarrow p$

The trivial tautology states that any statement is equivalent to itself. The proof is given by using a truth table.

p	$p \leftrightarrow p$
T	T
F	T

Since the last column contains all Ts, the result is proved.

5.5 The Law of Double Negation, $p \Leftrightarrow \sim(\sim p)$

In Chapter 2 we showed that $\sim(\sim p)$ could be replaced by p without affecting the truth values of the statements involved. We prove the tautology here by repeating the truth table for p and $\sim(\sim p)$ given in Chapter 2.

p	$\sim p$	$\sim(\sim p)$	$p \leftrightarrow \sim(\sim p)$
T	F	T	T
F	T	F	T

LAW OF DOUBLE NEGATION

$$p \leftrightarrow [\sim(\sim p)]$$

EXAMPLE 4

Let p: It is conceivable.

It is conceivable. ↔ It is not inconceivable. □

5.6 The Law of the Excluded Middle, $p \vee (\sim p)$

This tautology says that any given statement is either true or false. The proof is given by means of the following truth table.

p	$\sim p$	$p \vee (\sim p)$
T	F	T
F	T	T

EXAMPLE 5

Let p: Otto is guilty

Otto is guilty, or Otto is not guilty. □

LAW OF THE MIDDLE

$$p \vee \sim p$$

5.7 Direct Reasoning, $[(p \rightarrow q) \wedge p] \Rightarrow q$

This reasoning is also sometimes called *modus ponens,* the *law of detachment,* or reasoning by *assuming the antecedent.* Since this tautology is a little more involved, we will begin by considering the following example:

$p \rightarrow q$	If Hal receives an A on the final, then he passes the course.
p	Hal receives an A on the final.
$\therefore q$	Therefore Hal passes the course.

Recall from geometry that $\therefore$ is a symbol for the word *therefore*. *This argument consists of two* ***premises***, or **hypotheses**, and a **conclusion**; the argument is valid if

$$[(p \rightarrow q) \wedge p] \rightarrow q$$

is a tautology. By means of Table 12, we can see that the argument is valid.

Table 12
PROOF OF THE LAW OF DIRECT REASONING

p	q	$p \rightarrow q$	$(p \rightarrow q) \wedge p$	$[(p \rightarrow q) \wedge p] \rightarrow q$
T	T	T	T	T
T	F	F	F	T
F	T	T	F	T
F	F	T	F	T

The pattern of direct reasoning is very important and should be memorized. It is illustrated as follows.

DIRECT REASONING

Major premise:	$p \rightarrow q$
Minor premise:	p
Conclusion:	$\therefore\ q$

EXAMPLE 6

If Boris plays chess, then he is intelligent.
Boris plays chess.
Therefore Boris is intelligent.

If Judy is a logical person, then she will understand this example.
Judy is a logical person.
Therefore Judy understands this example.

We say that these arguments are *valid*, since we recognize them as examples of direct reasoning. □

5.8 Indirect Reasoning, $[(p \rightarrow q) \wedge (\sim q)] \Rightarrow (\sim p)$

This reasoning is also sometimes called *modus tollens* or reasoning by *denying the consequent*. The following example illustrates what we call indirect reasoning.

$p \rightarrow q$	If Hal receives an A on the final, then he passes the course.
$\sim q$	Hal does not pass the course.
$\therefore \sim p$	Therefore Hal does not receive an A on the final.

We can reason as follows: $\sim p$ must be true, for, if it were false, then p would be true; but if p is true, then q is true; this is impossible because of the minor premise. The symbolic form of the argument should be memorized.

INDIRECT REASONING

Major premise:	$p \to q$
Minor premise:	$\sim q$
Conclusion:	$\therefore \sim p$

We can prove the tautology by considering the truth table (see Table 13.)

Table 13
PROOF OF THE LAW OF INDIRECT REASONING

p	q	$(p \to q)$	$\sim q$	$(p \to q) \wedge \sim q$	$\sim p$	$[(p \to q) \wedge \sim q] \to \sim p$
T	T	T	F	F	F	T
T	F	F	T	F	F	T
F	T	T	F	F	T	T
F	F	T	T	T	T	T

EXAMPLE 7

If the cat takes the rat, then the rat takes the cheese.
The rat does not take the cheese.
Therefore the cat does not take the rat.

If Hal receives an A on that test, then I am Napoleon.
I am not Napoleon.
Therefore Hal does not receive an A on that test. □

5.9 Law of Transitivity, $[(p \to q) \wedge (q \to r)] \Rightarrow (p \to r)$

This tautology allows us to consider longer and more complicated chains of reasoning. The proof is left for the reader.

LAW OF TRANSITIVITY

$p \to q$
$q \to r$
$\therefore p \to r$

EXAMPLE 8

Let p: Linda attends class.
q: Linda will pass the course.
r: Linda will graduate.

If Linda attends class, then she will pass the course.
If Linda passes the course, then she will graduate.
Therefore, if Linda attends class, then she will graduate. □

Transitivity can be extended so that a chain of several if–then statements are connected together. For example, we could continue:

If Linda graduates, then she will get a good job.
If Linda gets a good job, then she will meet the right people.
If Linda meets the right people, then she will become well known.
Therefore, if Linda attends class, then she will become well known.

Recall that the king's famous cry in the climactic scene of *Richard III.*

"For Want of a Nail a Kingdom Was Lost," used this type of reasoning to conclude: "If I lose a nail, then I will lose a kingdom."

5.10 Law of Contraposition, $(p \rightarrow q) \Leftrightarrow (\sim q \rightarrow \sim p)$

In Section 3.2 we showed that a statement could always be replaced by its contrapositive without affecting its truth or falsity. (The truth values are shown in Table 8, on page 19.)

In the following symbolic statement, we can replace any entry in one column with the corresponding entry in the other column and not affect the truth or falsity of the argument.

LAW OF CONTRAPOSITION

Column 1	*Column 2*
$p \rightarrow q$	$\sim q \rightarrow \sim p$
$p \rightarrow \sim q$	$q \rightarrow \sim p$
$\sim p \rightarrow q$	$\sim q \rightarrow p$
$\sim p \rightarrow \sim q$	$q \rightarrow p$

5.11 Disjunctive Syllogism, $[(p \vee q) \wedge (\sim p)] \Rightarrow q$

Suppose you know that $p \vee q$ is true. Furthermore, you subsequently find out that p is false (that is, $\sim p$ is true). What can you say about q? This question is answered by what is called the disjunctive syllogism.

DISJUNCTIVE SYLLOGISM

$$\begin{array}{l} p \vee q \\ \underline{\sim p \quad\quad} \\ \therefore\ q \end{array}$$

This can be proved by using a truth table (see Table 14).

Table 14
PROOF OF THE DISJUNCTIVE SYLLOGISM

p	q	$p \vee q$	$\sim p$	$(p \vee q) \wedge (\sim p)$	$[(p \vee q) \wedge (\sim p)] \rightarrow q$
T	T	T	F	F	T
T	F	T	F	F	T
F	T	T	T	T	T
F	F	F	T	F	T

EXAMPLE 9

Let p: The big prize is behind door 1.
q: The big prize is behind door 2.

The big prize is behind door 1 or door 2.
The big prize is not behind door 1.
Therefore the big prize is behind door 2. □

5.12 PROBLEM SET 5

1. List the trivial tautology, and explain it in your own words.

2. List the law of double negation, and explain it in your own words.

3. List the law of the excluded middle, and explain it in your own words.

4. List direct reasoning, and explain it in your own words.

5. List indirect reasoning, and explain it in your own words.

6. List the law of transitivity, and explain it in your own words.

7. List the law of contraposition, and explain it in your own words.

8. List the disjunctive syllogism, and explain it in your own words.

Each example in Problems 9–33 illustrates one of the tautologies discussed in this chapter. Name the tautology that applies.

9. If Otto goes to church, then he sings hymns. Therefore, if Otto does not sing hymns, then he does not go to church.

10. A rose is a rose.

11. Either I am going or I am not going.

12. If 2 divides a positive integer N, then N is even. N is not even. Therefore 2 does not divide N.

13. "Nobody doesn't like Sarah Lee"; therefore "everybody likes Sarah Lee."

14. If Stanley the student riots, then he does not learn mathematics. Stanley learns mathematics, therefore he does not riot.

15. $x > 0$ or $x > 0$.

16. If it is not impossible, then it is possible.

17. If Stan is intelligent, then he is aware of some social injustices in our society. If Stan is aware of some social injustices in our society, then he has an obligation to try to correct those injustices. Therefore, if Stan is intelligent, then he has an obligation to try to correct social injustices.

18. If Stan studies hard, then he gets an A. Stan studies hard. Therefore Stan gets an A.

19. A good deal is a good deal.

20. I'm going to join the mathematics club or go to work. I don't join the mathematics club. Therefore I'll go to work.

21. If Hank has a proof of Goldbach's conjecture, then I am a monkey's uncle. I am not a monkey's uncle. Therefore Hank does not have proof of Goldbach's conjecture.

22. Bob mows the lawn on Friday or Saturday. Bob does not mow the lawn on Friday. Therefore Bob mows the lawn on Saturday.

23. If the car broke down, then we did not go on vacation. Therefore, if we went on vacation, then the car did not break down.

24. If a^2 is even, then a must be even. a is odd. (Assume that "a is odd" = "a is not even.") Therefore a^2 is odd.

25. The treasure is in the north corner or in the eastern sector. The treasure is not in the eastern sector. Therefore the treasure is in the north corner.

26. If it is not infeasible, then it is feasible.

27. If it is a snark, then it is a fribble. If it is a fribble, then it is ugly. Therefore, if it is a snark, then it is ugly.

28. If it is a snark, then it is a fribble. Therefore, if it is not a fribble, then it is not a snark.

29. If I inherit \$1,000, I will buy Charlie a cookie. I inherit \$1,000. Therefore I will buy Charlie a cookie.

30. If the quick brown fox jumps over the lazy dog, then he trips. Therefore, if the quick brown fox does not trip, then he doesn't jump over the lazy dog.

31. Man is just or unjust.

32. Enough is enough.

33. John or Gary is in charge. Gary is not in charge. Therefore John is in charge.

Make a truth table for Problems 34–42. These are well-known laws of logic.

34. Law of transitivity

35. De Morgan's law: $\sim(p \wedge q) \Leftrightarrow [\sim p \vee (\sim q)]$

36. De Morgan's law: $\sim(p \vee q) \Leftrightarrow [\sim p \wedge (\sim q)]$

37. Law of contradiction: $\sim[p \wedge (\sim p)]$

38. Law of implication and disjunction: $(p \rightarrow q) \Leftrightarrow (\sim p \vee q)$

39. Idempotent property: $(p \wedge p) \Leftrightarrow p$

40. Idempotent property: $(p \vee p) \Leftrightarrow p$

41. Reduction law: $[p \vee (p \wedge q)] \Leftrightarrow p$

42. Reduction law: $[p \wedge (p \vee q)] \Leftrightarrow p$

CHAPTER 6 LOGICAL FALLACIES

6.1 Validity

Symbolic logic is concerned with analyzing *valid arguments;* it cannot tell us whether the information in a hypothesis is true or false. The words *valid* and *invalid* refer to the argument form, not to the truth or falsity of the propositions. A valid argument asserts that the conclusion follows from the hypothesis according to one or more tautologies. In the last chapter we discussed some of the most common tautologies.

We can be sure that, *if* what we start with is true (or is assumed to be true), then, by knowing how to apply the tautologies, we will reach correct conclusions. That is, it is impossible to start with true premises and end up with a false conclusion by using the proper rules of logic. The important point to remember is that the truth or falsity of the premises and conclusion does not determine the **validity** of the argument.

Sometimes *invalid arguments* are given. In the remainder of this chapter, we will discuss some common invalid arguments, or *logical fallacies.*

6.2 Fallacy of Assuming the Consequent

EXAMPLE 1 Test the following argument for validity.

If a person reads the *Times,* then she is well informed.
This person is well informed.
Therefore this person reads the *Times.*

Solution This argument has the following form:

$$\begin{array}{l} p \rightarrow q \\ \underline{q \qquad\qquad} \\ \therefore p \end{array}$$

By considering the associated truth table, we can test the validity of our argument.

p	q	$p \rightarrow q$	$(p \rightarrow q) \wedge q$	$[(p \rightarrow q) \wedge q] \rightarrow p$
T	T	T	T	T
T	F	F	F	T
F	T	T	T	F
F	F	T	F	T

We see the result is not always true; thus the argument is invalid. □

If $p \to q$ were replaced by $q \to p$, the argument in the preceding example would become valid. That is, the argument would be valid if the direct statement and the converse had the same truth values, which in general they do not. For this reason the argument is sometimes called the **fallacy of the converse.**

We can often show that a given argument is invalid by finding a **counterexample.** In the preceding example we found a counterexample by looking at the truth table. The entry in the third row is false, so the argument can be shown to be false in the case in which p is false and q is true. In terms of this example, a person could never see the *Times* (p false) and still be well informed by reading the *Tribune* (q true).

EXAMPLE 2 Test the validity of the following argument.

If a person is a dope addict, then he smokes pot.
This person smokes pot.
___
Therefore this person is a dope addict.

Solution Since this argument is of the same form as the first example, we see that it is a case of invalid reasoning. □

6.3 Fallacy of Denying the Antecedent

Consider the following argument:

If a person reads the *Times,* then he is well informed.
This person does not read the *Times.*
___
Therefore this person is not well informed.

As we have seen, a person who reads the *Tribune* might also be well informed. This line of reasoning is called the **fallacy of denying the antecedent** (sometimes also called the **fallacy of the inverse**). A truth table for

$$[(p \to q) \land (\sim p)] \to (\sim q)$$

shows that the fallacy of denying the antecedent is not valid. (The truth table is left as a problem.) □

EXAMPLE 3 Test the validity of the following argument.

If a person goes to college, he will make a lot of money.
You do not go to college.
___
You will not make a lot of money.

Solution This argument has the following form:

$$\begin{array}{l} p \to q \\ \sim p \\ \hline \therefore \sim q \end{array}$$

We recognize this reasoning as the fallacy of denying the antecedent. □

6.4 False Chain Pattern

In certain areas of this country, it is thought that thunderstorms cause milk to sour. This belief is an example of the fallacy we will call the **false chain pattern**. It can be shown as follows:

Hot, humid weather favors thunderstorms.
Hot, humid weather favors bacterial growth, which causes milk to sour.
____
Therefore thunderstorms cause milk to sour.

The false chain pattern is illustrated by

$$\begin{array}{l} p \rightarrow q \\ p \rightarrow r \\ \hline \therefore\ q \rightarrow r \end{array}$$

and can be proved invalid by construction of an appropriate truth table.

The three types of common fallacies that we have discussed can be summarized as follows:

Fallacy of Assuming the Consequent	*Fallacy of Denying the Antecedent*	*False Chain Pattern*
$p \rightarrow q$	$p \rightarrow q$	$p \rightarrow q$
q	$\sim p$	$p \rightarrow r$
$\therefore\ p$	$\therefore\ \sim q$	$\therefore\ q \rightarrow r$

Study these fallacies, and notice how they differ from the tautologies of the previous chapter.

6.5 PROBLEM SET 6

1. By constructing a truth table, show that
$$[(p \rightarrow q) \wedge (p \rightarrow r)] \rightarrow (q \rightarrow r)$$
is an invalid argument. What is the name of this fallacy?

2. Prove that $[(p \rightarrow q) \wedge \sim p] \rightarrow \sim q$ is an invalid argument. What is the name of this fallacy?

In Problems 3–18, test the validity of each of the following arguments. If invalid, determine the error in reasoning.

3. If Congress appropriates the money, the project can be completed.
Congress appropriates the money.
____
Therefore the project can be completed.

4. If Alice drinks from the bottle marked "poison," she will become sick.
Alice drinks from a bottle that is not marked "poison."
____
Therefore she does not become sick.

5. Blue-chip stocks are safe investments.
Stocks that pay a high rate of interest are safe investments.
Therefore blue-chip stocks pay a high rate of interest.

6. If the United States builds an antiballistic system, our major cities will be protected from attack.
The United States does not build an antiballistic system.
Therefore our major cities will not be protected from attack.

7. If Al understands logic, then he enjoys this sort of problem.
Al does not understand logic.
Therefore Al does not enjoy this sort of problem.

8. If Al understands a problem, it is easy.
This problem is not easy.
Therefore Al does not understand this problem.

9. If I don't get a raise in pay, I will quit.
I don't get a raise in pay.
Therefore I will quit.

10. Mary has a little lamb or a big bear.
Mary does not have a big bear.
Therefore Mary has a little lamb.

11. If $2x - 4 = 0$, then $x = 2$.
$x \neq 2$
Therefore $2x - 4 \neq 0$.

12. If Todd eats Krinkles cereal, then he has "extra energy."
Todd has "extra energy."
Therefore Todd eats Krinkles cereal.

13. If Ron uses Slippery oil, then his car is in good running condition.
Ron's car is in good running condition.
Therefore Ron uses Slippery oil.

14. If you get a fill-up of gas, you will get a free car wash.
You get a fill-up of gas.
Therefore you will get a free car wash.

15. If the San Francisco 49ers lose, then the Dallas Cowboys win.
If the Dallas Cowboys win, then they will go to the Super Bowl.
Therefore, if the San Francisco 49ers lose, then the Dallas Cowboys will go to the Super Bowl.

16. If Missy uses Smiles toothpaste, then she has fewer cavities.
Therefore, if Missy has fewer cavities, then she uses Smiles toothpaste.

17. (Let x be an integer and y a nonzero integer for this argument.)
If Q is a rational number, then $Q = \frac{x}{y}$, where $\frac{x}{y}$ is a reduced fraction.
$Q \neq \frac{x}{y}$, where $\frac{x}{y}$ is a reduced fraction.
Therefore Q is not a rational number.

18. If the crime occurred after 4:00 A.M., then Smith could not have done it.
If the crime occurred on or before 4:00 A.M., then Jones could not have done it.
The crime involved two persons, if Jones did not commit the crime.

Therefore if Smith committed the crime, it involved two persons.

CHAPTER 7

QUANTIFIERS AND EULER CIRCLES

7.1 Universal and Existential Quantifiers

Up to now we have been discussing the logical inference of structure only of statements that are classified as true or false. However, in mathematics we sometimes need to consider three types of sentences: (1) true, (2) false, and (3) open. Examples of each of these types are given below.

1. True sentences
   a. $2 + 3 = 5$
   b. $6 < 9$
   c. George Washington was the first U.S. President.
2. False sentences
   a. $2 + 3 = 9$
   b. $5 < 2$
   c. Alfred E. Newman was the first U.S. President.
3. Open sentences
   a. $2x^2 + 3x = 5$
   b. $5y < 20$
   c. He was the first U. S. President.

We see that these open sentences may be true or false, depending on the replacements for x, y, and *he,* whereas true sentences must *always* be true and false sentences must *always* be false.

We now wish to apply our study of logic to open sentences. To do so we must restrict or *quantify* the variable (x, y, or *he* in our examples), by saying that it is true for all or some of its possible replacements.

For example, $2x + 3x = 5x$ is an open sentence that is true for all possible replacements of x; thus we change this open sentence into a true sentence by saying

For all x, $2x + 3x = 5x$.

EXAMPLE 1

Change each of the following open sentences into a true sentence.

1. $x + 2 = 2 + x$
2. Cockroaches are ugly.
3. Two right angles are congruent.
4. $x + y = y + x$

Solution

1. For all x, $x + 2 = 2 + x$.
2. All cockroaches are ugly.
3. Any two right angles are congruent.
4. For all x and for all y, $x + y = y + x$. □

Any quantifier of the form *for all, all, for each,* or *for every* is called a **universal quantifier** and is symbolized by an upside-down A:

1. $\forall x,\ x + 2 = 2 + x$ 2. $\forall x \forall y,\ x + y = y + x$

Sometimes we cannot change an open sentence into an equivalent statement by using a universal quantifier. For example, the open sentence $5y < 20$ is not true for all replacements of y. It is true for at least one y ($y = 1$, for example), however, although it may not be true for more than one y. To make the open sentence $5y < 20$ into a true sentence, we would say

There exists a y such that $5y < 20$.

EXAMPLE 2

Change each of the following open sentences into a true sentence.

1. $2x + 2 = 5$
2. $4z < 19$
3. He was the first U.S. President.
4. Some apples are red.

Solution

1. There exists an x such that $2x + 2 = 5$.
2. There exists a z such that $4z < 19$.
3. There exists a person who was the first U. S. President.
4. There exists at least one apple that is red. □

We will translate the word *some* to mean *at least one.* Quantifiers of the form *there exists, some,* or *there is at least one* are called **existential quantifiers** and are denoted by a backward E:

1. $\exists x$ such that $2x + 2 = 5$. 2. $\exists z$ such that $4z < 19$.

7.2 Truth Values of Quantified Sentences

We see that, in the case of an open sentence or of an open sentence that has been quantified, some universal set has been either stated or implied. By a *universal set* we mean a set that contains all the elements under consideration in a given discussion. In addition, there is a set of possible replacements for the variable. This set is called the *replacement set.* The truth values of these quantified sentences may vary as the replacements vary.

For example,

$$\forall x,\ x = 1$$

is a statement with a universal quantifier. Suppose we let the universal set be the set of all real numbers. Then we would say that our statement is false. If the universal set is the set $\{1\}$, however, the statement is true. Hence, the truth value of a statement with a universal quantifier depends on the universal set. This leads to the following definition.

TRUTH VALUE WITH UNIVERSAL QUANTIFIERS

A statement with a universal quantifier is true if and only if the replacement set of the variable is the same as the universal set for the problem.

On the other hand, when we use the existential quantifier, we mean that there is (at least) *one* replacement for the variable that makes the statement true. For example,

$$\exists x \text{ such that } x^2 - 1 = 0$$

where the universal set is the set of real numbers, is true, since we can find at least one replacement for x that makes the sentence true. However,

$$\exists x \text{ such that } x^2 + 1 = 0$$

where the universal set is the set of real numbers, is false, since there are *no* replacements for x that make the statement true.

TRUTH VALUE WITH EXISTENTIAL QUANTIFIER

A statement with an existential quantifier is true if and only if the replacement set of the variable is nonempty.

7.3 Introduction to Euler Circles

As we introduce quantifiers, the relationships between various statements may not be obvious. Thus we will seek a way to picture the relationships between various propositions. The Swiss mathematician Leonhard Euler (1707–1783) used circles or oval-shaped regions to illustrate the relationships between statements in logic; these diagrams have become known as *Euler circles.* They were also used by John Venn (1834–1923) and are sometimes called *Venn diagrams.*

Let's begin by considering the Euler circles for two propositions p and q. There are several possibilities, as shown in Figure 1.

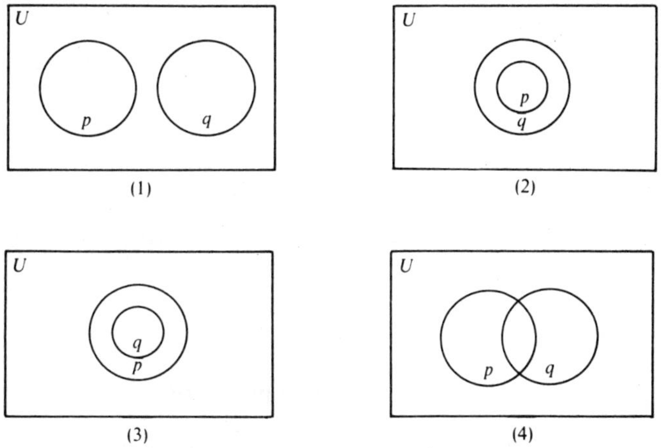

Figure 1 Possible relationships between two propositions p and q

Let the universal set be the set of all persons. Also let p: Persons who can learn mathematics, and q: Persons who are intelligent. Then, (1) of Figure 1 illustrates the situation in which no person who can learn mathematics is an intelligent person. In drawing (2) we see that "If a person can learn mathematics, then he is intelligent"; and drawing (3) illustrates the converse, "If a person is intelligent, then he can learn mathematics." These statements could also be written using the quantifier *all*; for example, drawing (3) can be written as "All intelligent persons can learn mathematics."

Suppose that we do not know the relationship between two propositions p and q. Then we draw a figure similar to drawing (4), as shown in Figure 2.

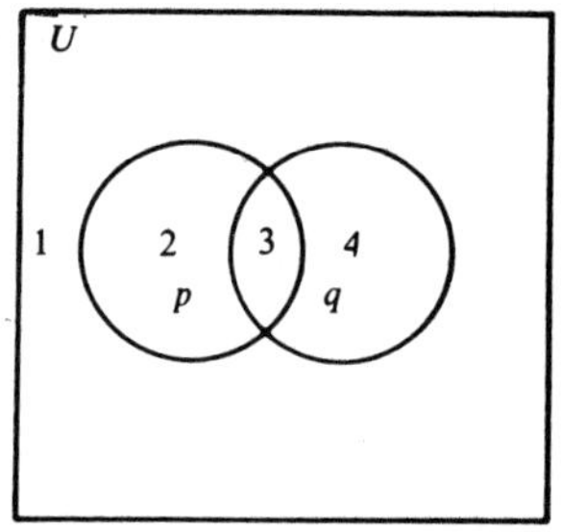

Figure 2 General Euler diagram for two propositions

Notice that the universe, represented by a rectangle, is divided into four regions. When we draw the diagram in this manner, we do not mean to imply that p and q are necessarily overlapping.

If all p are q, then region 2 is empty.
If all q are p, then region 4 is empty.
If $p \leftrightarrow q$, then regions 2 and 4 are both empty.
If no p is q, then region 3 is empty.

That is, when we draw overlapping propositions, any one of these possibilities might be the case.

7.4 Simple Connectives

If we wish to represent conjunction, disjunction, or negation, we begin with a general diagram. For one proposition:

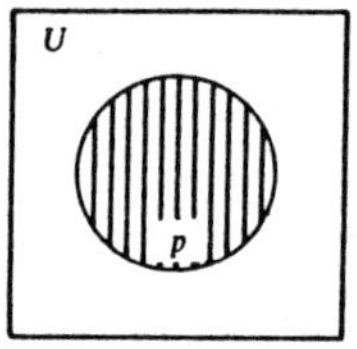

p

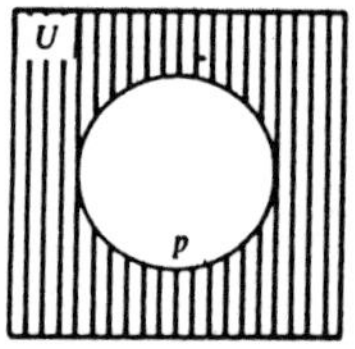

$\sim p$

For two propositions:

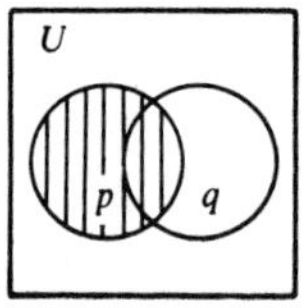

p

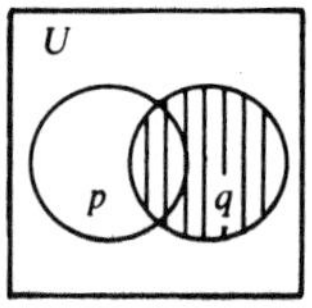

q

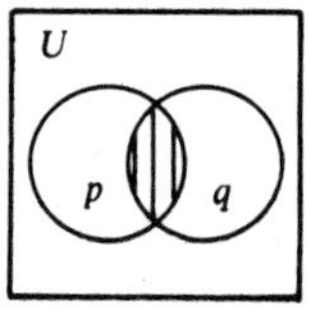

$p \wedge q$

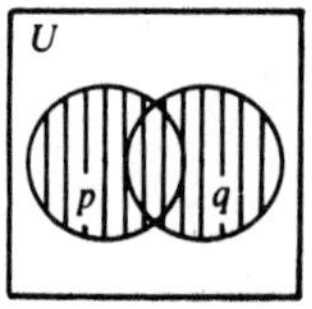

$p \vee q$

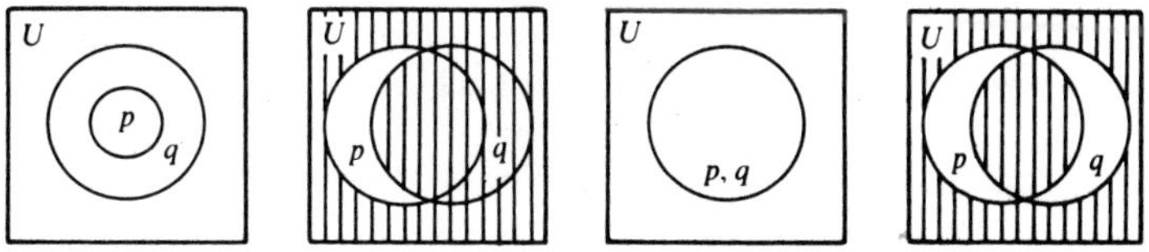

(1) (2) $p \to q$ (1) (2) $p \leftrightarrow q$

Notice the two representations for $p \to q$ and $p \leftrightarrow q$. Representation (1) is used if we *know* that $p \to q$ before we begin. On the other hand, we might not know how p and q are related before we begin. In this case two general sets are drawn, as shown in Figure 2. Then, if we later find that $p \to q$, the relationship would be shaded, as shown in representation (2).

As an example, let's use Euler circles to show direct reasoning. Recall that:

$$\begin{array}{l} p \to q \\ p \\ \hline \therefore q \end{array}$$

First we draw the diagram to illustrate the first premise, $p \to q$.

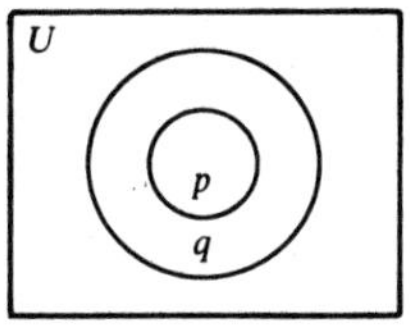

The second premise gives us p. We illustrate this by placing an x inside circle p.

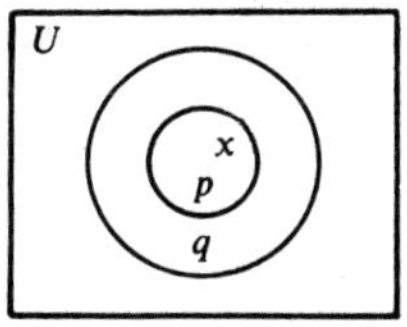

Now the conclusion q follows, since the diagram shows an x inside q. There is no possible way the figure can be drawn so the x is outside q.

EXAMPLE 3 Illustrate the law of transitivity using Euler circles.

Solution Recall that:

$$\begin{array}{l} p \to q \\ q \to r \\ \hline \therefore p \to r \end{array}$$

We draw circles for the premises:
(1) p inside q
(2) q inside r

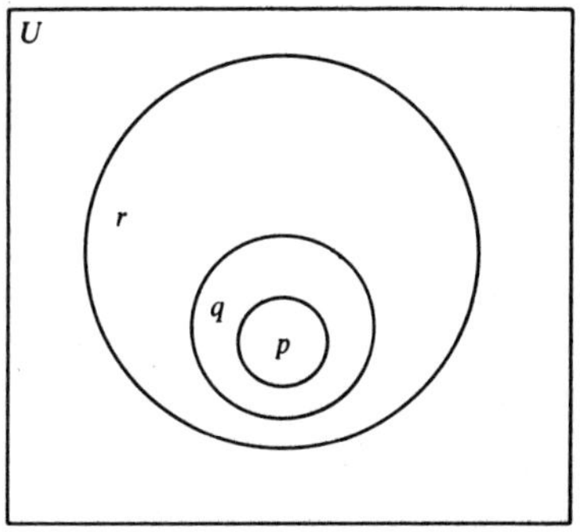

Now the conclusion $p \rightarrow r$ follows, since we see that we are *forced* to put p inside r. □

7.5 Logical Arguments

We will illustrate the method of using Euler circles to analyze some logical arguments.

EXAMPLE 4

Form a valid conclusion.

1. If you are good-looking, then you marry a millionaire.
2. You do not marry a millionaire

Solution

Let U: all people (U represents the universe)
g: good-looking people
m: people married to millionaires

Now the first premise gives us the following diagram:

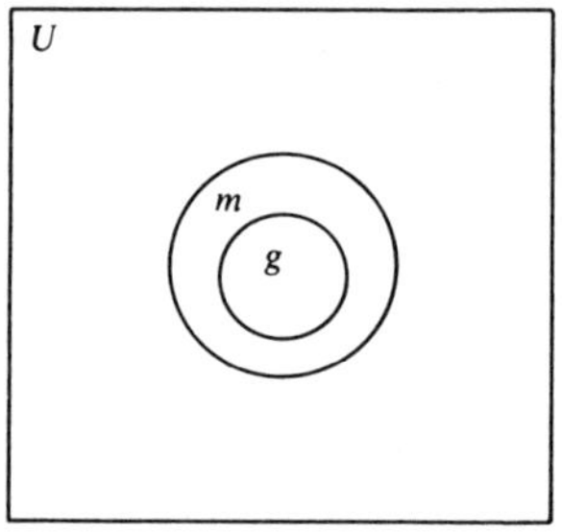

The second premise tells us to place an x representing "you" outside the set of millionaires. This *requires* that the x be outside the circle marked m.

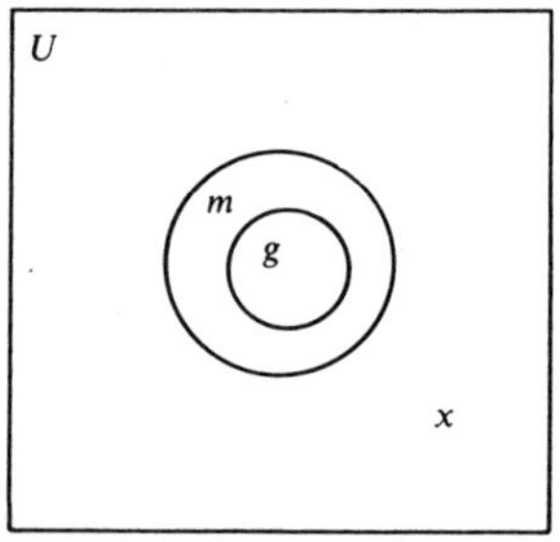

Therefore the conclusion "You are not good-looking" follows. □

EXAMPLE 5

Form a valid conclusion.*

1. All unripe fruit is unwholesome.
2. All these apples are wholesome.
3. No fruit grown in the shade is ripe.

Solution

Let U: fruit
r: ripe fruit
w: wholesome fruit
a: these apples
s: fruit grown in the shade

*From Lewis Carroll, *Symbolic Logic and the Game of Logic* (New York: Dover Publications, Inc., New York, 1958).

The first premise tells us to draw a circle w inside a circle r, since the sentence

$$\sim r \rightarrow (\sim w)$$

can be rewritten as

$w \rightarrow r$ *Law of Contraposition*

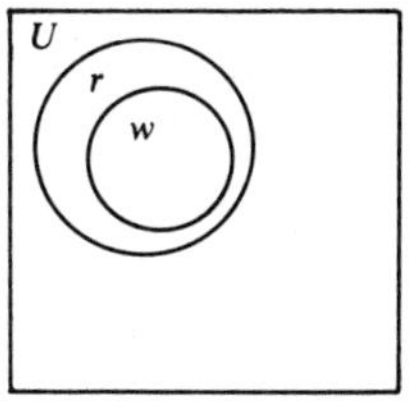

The second premise tells us to draw a circle a inside the circle w.

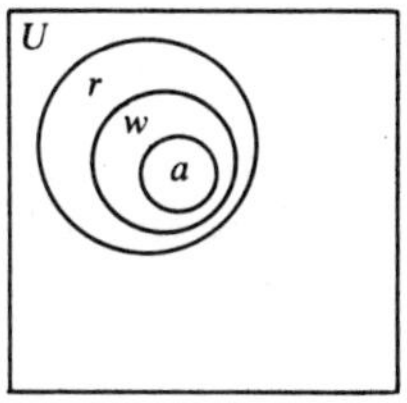

Now, the third premise forces us to draw the set s outside the set r.

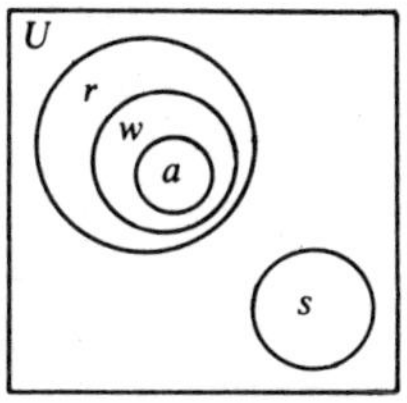

We see that more than one valid conclusion is possible. "All these apples were not grown in the shade" is a conclusion using all the premises. □

EXAMPLE 6

Test the validity of the following argument.

1. If you like potato chips, then you will like Krinkles.
2. You do not like potato chips.

Therefore you will not like Krinkles.

Solution

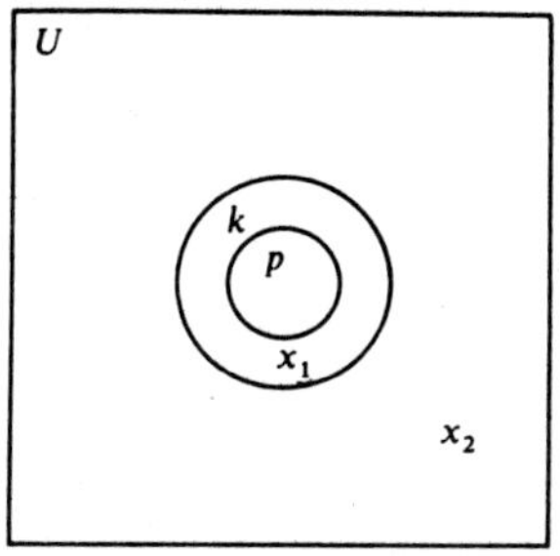

Let U: people
p: people who like potato chips
k: people who like Krinkles

We must draw p inside k, since all p are k (premise 1). However, if you are not in p, you may or may not be in k, as shown (see x_1 and x_2). Therefore the argument is *not valid*. □

7.6 PROBLEM SET 7

Determine the truth values of the following statements.

EXAMPLE 7

a. $\forall x, x + 2 = 10$
b. $\exists x$ such that $x + 2 = 10$
c. $\forall x, 3x + 10x = 13x$

Solution

a. False, since $x = 3$ makes the sentence false.
b. True, since there exists an x (namely $x = 8$) that makes the sentence true.
c. True, since all replacements for x make the sentence true. □

1. $\forall x, x + 3x = 4x$
2. $\forall x, x - 7 = 10$
3. $\exists x$ such that $3x - 4 = 11$
4. $\exists x$ such that $2x < 13$
5. Every integer is divisible by itself and 1.

Write each of the open sentences in Problems 6–10 as a true statement by using the most general quantifier.

6. $5x + 7 = 13$
7. $3y < 10$
8. $5x \cdot 1 = 5x$
9. $xy = yx$
10. $a(b + c) = ab + ac$

Write appropriate symbols for the universal or existential quantifiers used in Problems 11–15.

11. Some apples are rotten.
12. All women are beautiful.
13. There exist positive integers.
14. Some integers are negative.
15. All squares are rectangles.

Use Euler circles to represent the expressions in Problems 16–25.

16. $\sim p \wedge q$
17. $\sim p \wedge (\sim q)$
18. $p \rightarrow (\sim q)$
19. $p \vee (\sim q)$
20. $\sim p \vee (\sim q)$
21. $q \rightarrow (\sim p)$
22. $p \vee (q \vee r)$
23. $(p \vee q) \vee r$
24. $p \vee (q \wedge r)$
25. $(p \vee q) \wedge (p \vee r)$

26. Use Euler circles to show indirect reasoning.

In Problems 27–33 use Euler circles to check the validity of each argument.

27. All mathematicians are eccentrics.
All eccentrics are rich.
Therefore all mathematicians are rich.

28. Some beautiful women are blonde.
Blondes have more fun.
Therefore some beautiful women have more fun.

29. All bachelors are handsome.
Some bachelors do not drink lemonade.
Therefore some handsome men do not drink lemonade.

30. Dictionaries are useful.
Useful books are valuable.
Therefore dictionaries are valuable.

31. No professors are ignorant.
All ignorant people are vain.
Therefore no professors are vain.

32. No monkeys are soldiers.
All monkeys are mischievous.
Therefore some mischievous creatures are not soldiers.

33. All lions are fierce.
Some lions do not drink coffee.
Therefore some creatures that drink coffee are not fierce.

CHAPTER 8 THE NATURE OF PROOF

8.1 What Is Proof?

In mathematics we are often concerned with the logical proof of certain assertions. Any logical system must start with some undefined terms, definitions, and postulates or axioms. From there, other assertions can be made. These assertions, called theorems, must be proved using the rules of logic.

The steps we will use to prove logical arguments were introduced in Chapter 1:

1. Translate the sentences into symbolic form.
2. Apply one or more of the tautologies of Chapter 6.
3. After the argument is simplified through one or more of the fundamental tautologies, translate the result back into ordinary language.

In proving an argument, it is necessary to supply a reason for each step in the proof. The following examples provide a correct method for proving arguments. Notice that each premise of an argument is numbered for easy reference. If one of these premises is changed to an equivalent form, we will indicate it by a prime mark; thus, if premise 1 is changed to an equivalent form, we will denote it by $1'$.

8.2 Example of Logical Arguments

EXAMPLE 1 Form a valid conclusion using all the statements.*

1. If you are good-looking, then you marry a millionaire.
2. You do not marry a millionaire.

Solution The pattern can be written as follows:

1. $g \to m$	where	g:	You are good-looking.
2. $\sim m$		m:	You marry a millionaire.
$\therefore\ \sim g$	*Indirect reasoning*		

Conclusion: *You are not good-looking.* □

EXAMPLE 2 Form a valid conclusion using all the statements.

1. If we go on a picnic, then we pack a lunch.
2. If we pack a lunch, then we buy bread.
3. If we buy bread, then we go to the store.

---

*We solved this problem in the last chapter by using Euler circles. Now we consider an alternative method of solution.

Solution

First we change the argument to symbolic form:

1.	$p \to \ell$	where p:	We go on a picnic.
2.	$\ell \to b$	ℓ:	We pack a lunch.
3.	$b \to s$	b:	We buy bread.
		s:	We go to the store.

Next we simplify the argument, using the tautologies of Chapter 6:

1. $p \to \ell$
2. $\underline{\ell \to b}$
   $p \to b$ *Transitive*
3. $\underline{b \to s}$
   $p \to s$ *Transitive*

This could also be expressed as a chain:

$p \to l \to b \to s \qquad \therefore p \to s$

Finally, we translate the conclusion back into English:

If we go on a picnic, then we go to the store. □

EXAMPLE 3

Form a valid conclusion using all the statements.

1. If I receive a check for $500, then we will go on vacation.
2. If the car breaks down, then we will not go on vacation.
3. The car breaks down.

Solution

Step 1: Translate

1.	$c \to v$	where c:	I receive a check for $500.
2.	$b \to (\sim v)$	v:	We will go on vacation.
3.	b	b:	The car breaks down.

(Do not assume $c \to b$ or $b \to c$; this is a false chain pattern.)

Step 2: Simplify

Rearrange the premises, and use the law of contraposition to replace (1) with (1′). Here 1′ is $\sim v \to \sim c$.

2. $b \to (\sim v)$
1′. $\underline{\sim v \to (\sim c)}$
   $\therefore b \to (\sim c)$ *Transitive*
3. $\underline{b}$
   $\therefore \sim c$ *Direct reasoning*

Step 3: Translate

Conclusion: *I do not receive a check for $500.* □

EXAMPLE 4

Form a valid conclusion using all the statements.*

1. All unripe fruit is unwholesome.
2. All these apples are wholesome.
3. No fruit grown in the shade is ripe.

*From Lewis Carroll, *Symbolic Logic and The Game of Logic* (New York: Dover Publications, 1958).

Solution

Step 1:

1.	$\sim r \rightarrow (\sim w)$	where r:	ripe fruit
2.	$a \rightarrow w$	w:	wholesome fruit
3.	$s \rightarrow (\sim r)$	a:	these apples
		s:	fruit grown in the shade

Note: The sentence "No p is q" is translated as $p \rightarrow (\sim q)$.

Step 2:
Let (1′), $w \rightarrow r$, replace (1) by the law of contraposition.
Let (3′), $r \rightarrow \sim s$, replace (3) by the law of contraposition.
Rearranging the premises, we have:

2. $a \rightarrow w$
1′ $w \rightarrow r$

$\therefore a \rightarrow r$ *Transitive*

3′ $r \rightarrow (\sim s)$

$a \rightarrow (\sim s)$ *Transitive*

Step 3:
Conclusion: *All these apples were not grown in the shade.* If we assume that $\sim s$ = grown in the sun, then the conclusion can be stated more simply: *All these apples were grown in the sun.* □

8.3 PROBLEM SET 8

Form a valid conclusion, using all the statements for each argument. Give reasons.

1. If you learn mathematics, then you are intelligent.
If you are intelligent, then you understand women.

2. If I am idle, then I become lazy.
I am idle.

3. If we go to the concert, then we are cultured.
We are not cultured.

4. If you climb the highest mountain, then you feel great.
If you feel great, then you are happy.

5. All Trebbles are Frebbles.
All Frebbles are expensive.

6. $a = 0$ or $b = 0$
$a \neq 0$

7. If $a \cdot b = 0$, then $a = 0$ or $b = 0$.
$a \cdot b = 0$

8. If I buy my wife a purse, then I buy her shoes.
If I buy my wife shoes, then I buy her a dress.
If I buy my wife a dress, then she goes to the beauty parlor.
If my wife goes to the beauty parlor, then I take her out for dinner.
If I take my wife out for dinner, then my mother-in-law

babysits.
If my mother-in-law babysits, then my father-in-law goes to the races.
If my father-in-law goes to the races, it costs him $5,000.

9. If we interfere with the publication of false information, we are guilty of suppressing the freedom of others.
We are not guilty of suppressing the freedom of others.

10. If 2 divides a positive integer N and if N is greater than 2, then N is not a prime number.
N is a prime number.

11. Babies are illogical.
Nobody is despised who can manage a crocodile.
Illogical persons are despised.

12. All hummingbirds are richly colored.
No large birds live on honey.
Birds that do not live on honey are dull in color.

13. No ducks waltz.
No officers ever decline to waltz.
All my poultry are ducks.

14. Everyone who is sane can do logic.
No lunatics are fit to serve on a jury.
None of your sons can do logic.

15. If the government awards the contract to Airfirst Aircraft Company, then Senator Firstair stands to earn a great deal of money.
If Airsecond Aircraft Company does not suffer financial setbacks, then Senator Firstair does not stand to earn a great deal of money.
The government awards the contract to Airfirst Aircraft Company.

16. If you go to college, then you get a good job.
If you get a good job, then you make a lot of money.
If you do not obey the law, then you do not make a lot of money.
You go to college.

17. No kitten that loves fish is unteachable.
No kitten without a tail will play with a gorilla.
Kittens with whiskers always love fish.
No teachable kitten has green eyes.
No kittens have tails unless they have whiskers.

18. When I work a logic example without grumbling, you may be sure it is one that I can understand.
These problems are not arranged in regular order, like the examples I am used to.
No easy example ever makes my head ache.
I can't understand examples that are not arranged in regular order, like those I am used to.
I never grumble at an example unless it gives me a headache.

* CHAPTER 9

PUZZLE PROBLEMS IN LOGIC

9.1 An Alternative Method of Proof

Let us apply another method of proof to solving some logic puzzles that have been around for a long time but that nevertheless continue to challenge the lay reader.

EXAMPLE 1
Liar Problem

On the South Side, a member of the mob had just knocked over a store.
Since the boss had told them all to lay low, he was a bit mad. He decided to have a talk with the boys.

Alfie said "It was Bogie or Clyde."

Bogie said "Neither Fingers nor I did it."

Clyde said "Both of you are lying."

Dirty Dave said "No, one of them is lying; the other is telling the truth."

Fingers Mulligan said "No, Dirty Dave, that's not true."

Now, the boss knew that three of the mob were always truthful, but two would lie every time. Can he figure out who knocked over the store?

Solution

Let a: Alfie did it.
b: Bogie did it.
c: Clyde did it.
d: Dirty Dave did it.
f: Fingers Mulligan did it.

Translate the sentences into symbolic form:

Alfie said: $b \vee c$
Bogie said: $\sim f \wedge (\sim b)$
Clyde said: $[\sim(b \vee c)] \wedge \sim[\sim(f \vee b)]$
Dirty Dave said: $[(b \vee c) \wedge (f \vee b)] \vee \{\sim(b \vee c) \wedge [\sim(f \vee b)]\}$
Fingers said: $\sim\Big([(b \vee c) \wedge (f \vee b)] \vee \{\sim(b \vee c) \wedge [\sim(f \vee b)]\}\Big)$

Assume that a is true, and check all these statements. Next, assume that b is true, and check all the statements. Do the same for c, d, and

*Optional chapter

f. Only in one case are three truthful and two lying. Thus Clyde knocked over the store. □

Since this solution is rather lengthy, let's introduce an alternative method of solution. We use a matrix (a rectangular array) with vacant cells for all possible pairings of the elements in each set.

	a	*b*	*c*	*d*	*f*
a					
b					
c					
d					
f					

Let us assume that the *a* at the left means that we are making the assumption that Alfie did it. Then we place a 1 or a 0 in each position to indicate the truth or falsity, respectively, of each witness' statement with the assumption that person x (as listed at the left) is guilty.

Conclusion:	**Witness statements:**	*a*	*b*	*c*	*d*	*f*
Alfie did it:	*a*	0	1	0	1	0
Bogie did it:	*b*	1	0	0	1	0
Clyde did it:	*c*	1	1	0	0	1
Dave did it:	*d*	0	1	0	1	0
Fingers did it:	*f*	0	0	1	0	1

Next we assume that Bogie did it, and we fill in the second row of the matrix. We continue until the matrix is complete, as shown. Now if we take the boss' statement as the major premise, we see that in only one case are there three truthful mobsters and two liars. Thus we can say that Clyde knocked off the store. □

EXAMPLE 2
Prisoner Problem

During an ancient war three prisoners were brought into a room. In the room was a large box containing three white hats and two black hats. Each man was blindfolded, and one of the hats was placed on his head. The men were lined up, one behind the other, facing the wall. The blindfold of the man farthest from the wall was removed, and he was permitted to look at the hats of the two men in front of him. If he knew (not guessed) the color of the hat on his head, he would be freed. However, he was unable to tell. The blindfold was then taken from the head of the next man, who could see only the hat of the one man in front of him. This man had the same chance for freedom, but he, too, was unable to tell the color of his hat. The remaining man then told the guards the color of the hat he was wearing and was released. What color hat was he wearing, and how did he know?

Solution

The situation looks like this:

WALL ▯ *A* *B* *C*

1. What did man *C* see? There are four possibilities:

	I	II	III	IV
A	white	white	black	black
B	white	black	white	black

If he had seen the fourth possibility, he would have known that he had on a white hat. (Why?) He did not know, so it must be possibility I, II, or III.

2. What did man *B* see? If he had seen a black hat, then he would have know, that his hat was white. (Why?) This rules out case III.

3. Since both of the remaining possibilities, I and II, have a white hat on the front man, man *A* *knew* he had a white hat. □

EXAMPLE 3
Odd Ball Problem

You are given nine steel balls of the same size and color. One of the nine balls is slightly heavier in weight; the others all weigh the same. Using a two-pan balance, what is the minimum number of weighings necessary to find the ball of different weight?

Solution

Two weighings.

1. Divide the 9 steel balls into 3 groups of 3. First weighing: balance 3 balls against 3 balls.
    a. The weighing either balances or doesn't balance (law of excluded middle).
    b. If it balances, then the heavier one is in the group not weighed. If it doesn't balance, then take the group with the heavier ball.
2. Divide the 3 steel balls from the heavier group. Second weighing: balance 1 ball against 1 ball.
    a. The weighing either balances or doesn't balance. (Why?)
    b. If it doesn't balance, then the heavier ball is the one that tips the scale. If it does balance, then the heavier ball is the one not weighed. □

EXAMPLE 4
Reporter's Error Problem

Emor D. Nilap, a new reporter who was a little backward, was sent to cover a billiards tournament. Since he wanted to do a good job, he rounded up some human-interest facts to make his story a little more interesting. He gathered the following information:

1. The men competing were Pat, Milt, Dick, Joe, and Karl.
2. Milt had once beaten the winner at tennis.

3. Pat and Karl frequently played cards together.
4. Joe's finish ahead of Karl was totally unexpected.
5. The man who finished fourth left the tournament after his match and did not see the last two games.
6. The winner had not met the man who came in fifth prior to the day of the tournament.
7. The winner and the runner-up had never met until Joe introduced them just before the final game.

When Emor returned to his office, he found that he had forgotten the order in which the men had finished but he was able to figure it out from the information provided. Can you figure it out?

Solution

From (1) we are able to make the following matrix of the possible solutions:

Pat	Milt	Dick	Joe	Karl
1	1	1	1	1
2	2	2	2	2
3	3	3	3	3
4	4	4	4	4
5	5	5	5	5

a. We are given seven premises. We begin by the process of elimination; that is, we cross out those positions that it would be impossible for the men to occupy. The order in which we work the problem is not unique.

b. From (7), Joe is not the winner or the runner–up. Thus we cross out 1 and 2 under Joe's name and label those places *b*, as shown on the next page.

c. By (5), Joe did not finish fourth.

d. Joe finished ahead of Karl, by (4); therefore Karl was not first, second, or third. By the same premise, Joe could not have been last. (Why?)

e. Therefore Joe finished third, which means that Pat, Milt, and Dick are not in third place.

f. By (2), Milt was not the winner.

g. By (7), Milt was not the runner-up.

h. By (6), Milt was not fifth.

i. Therefore Milt finished fourth, which means that Pat, Dick, and Karl are not in fourth place.

j. Therefore Karl finished fifth, which means that the others did not finish fifth.

k. By (6) and the fact that Karl finished last, along with (3), we see that Pat could not have finished first.

ℓ. Therefore Pat is second, which means that Dick was not second.

m. Therefore Dick finished first, and the problem is finished.

Pat	Milt	Dick	Joe	Karl
$\not{1}^k$	$\not{1}^f$	$\boxed{1}^m$	$\not{1}^b$	$\not{1}^d$
$\boxed{2}^\ell$	$\not{2}^g$	$\not{2}^\ell$	$\not{2}^b$	$\not{2}^d$
$\not{3}^e$	$\not{3}^e$	$\not{3}^e$	$\boxed{3}^e$	$\not{3}^d$
$\not{4}^i$	$\boxed{4}^i$	$\not{4}^i$	$\not{4}^c$	$\not{4}^i$
$\not{5}^j$	$\not{5}^h$	$\not{5}^j$	$\not{5}^d$	$\boxed{5}^j$

□

EXAMPLE 5
Turkey Problem

A man I know once owned a number of turkeys. One day, one of his gobblers flew over the man's fence and laid an egg on a neighbor's property. To whom did the egg belong — to the man who owned the gobbler, to the gobbler, or to the neighbor?

Solution

A turkey gobbler could not lay an egg. You must be careful to use common sense along with the rules of logic when answering puzzle problems. □

9.2 PROBLEM SET 9

1. *The Marble Players.* Four boys were playing marbles. Their names were Gary, Harry, Iggy, and Jack. One of the boys had 9 marbles, another had 15, and the other two each had 12. Their ages were 3, 10, 17, and 18, but not respectively. Gary shot before Harry and Jack. Jack was older than the boy with 15 marbles. Harry had fewer than 15 marbles. Jack shot before Harry. Iggy shot after Harry. If Iggy was 10 years old, he did not have 15 marbles. Gary and Jack together had an even number of marbles. The youngest boy was not the one with 15 marbles. If Harry had 12 marbles, he wasn't the youngest. The 10-year-old shot after the 17-year-old. In what order did they shoot, and how old was each boy?

2. *Bear Problem.* A fox, hunting for a morsel of food, spotted a huge bear about a hundred yards due east of him. Before the hunter could become the hunted, the crafty fox ran due north for a hundred yards but then realized the bear had not noticed him. Thus he stopped and remained hidden. At this point the bear was due south of the fox. What was the color of the bear?

3. *Whodunit?* Daniel Kilraine was killed on a lonely road, two miles from Pontiac, at 3:30 A.M., March 3, 1967. Otto, Curly, Slim, Mickey, and The Kid were arrested a week later in Detroit and questioned. Each of the five made four statements, three of which were true and one of which was false. One of these men killed Kilraine. Whodunit? Their statements were:*

Otto: "I was in Chicago when Kilraine was murdered. I never killed anyone. The Kid is the guilty man. Mickey and I are pals."

---

*Reprinted with permission of The Macmillan Company from *Introduction of Logic* (2d ed.), by Irving Copi. Copyright © 1961 by The Macmillan Company.

Curly: "I did not kill Kilraine. I never owned a revolver in my life. The Kid knows me. I was in Detroit the night of March 3rd."

Slim: "Curly lied when he said he never owned a revolver. The murder was committed on March 3rd. One of us is guilty. Otto was in Chicago at the time."

Mickey: "I did not kill Kilraine. The Kid has never been in Pontiac. I never saw Otto before. Curly was in Detroit with me on the night of March 3rd."

The Kid: "I did not kill Kilraine. I have never been in Pontiac. I never saw Curly before. Otto lied when he said I am guilty."

4. *Coin Problem.* Suppose you are given 12 coins, one of which is counterfeit and weighs a little more or less than the real coins. Using a two-pan balance scale, what is the minimum number of weighings necessary to find the counterfeit coin? (If you are interested in a general solution to this type of problem, see T. H. O'Beirne's book, *Puzzles and Paradoxes,* Oxford University Press, New York, 1965, Chapters 2 and 3.)

5. *Baseball Problem.* Nine men play the positions on a baseball team. Their names are Brown, White, Adams, Miller, Green, Hunter, Knight, Smith, and Jones. Determine from the following information the position played by each man.

a. Brown and Smith each won $10 playing poker with the pitcher.

b. Hunter is taller than Knight and shorter than White, but all three weigh more than the first baseman.

c. The third baseman lives across the corridor from Jones in the same apartment house.

d. Miller and the outfielders play bridge in their spare time.

e. White, Miller, Brown, the right fielder, and the center fielder are bachelors, and the rest are married.

f. Of Adams and Knight, one plays an outfield position.

g. The right fielder is shorter than the center fielder.

h. The third baseman is a brother of the pitcher's wife.

i. Green is taller than the infielders, the pitcher, and the catcher except for Jones, Smith, and Adams.

j. The second baseman beat Jones, Brown, Hunter, and the catcher at cards.

k. The third baseman, the shortstop, and Hunter each made $150 speculating in General Motors stock.

ℓ. The second baseman is engaged to Miller's sister.

m. Adams lives in the same apartment house as his own sister but dislikes the catcher.

n. Adams, Brown, and the shortstop each lost $200 speculating in grain.

o. The catcher has three daughters, the third baseman has two sons, but Green is being sued for divorce.

*CHAPTER 10 LOGIC CIRCUITS

10.1 Schematics

In this section we will apply logic to the theory of switching circuits. For our purposes we will assume that all switches are two-state devices; that is, they are either *on* or *off*. Schematically a switch is represented as follows:

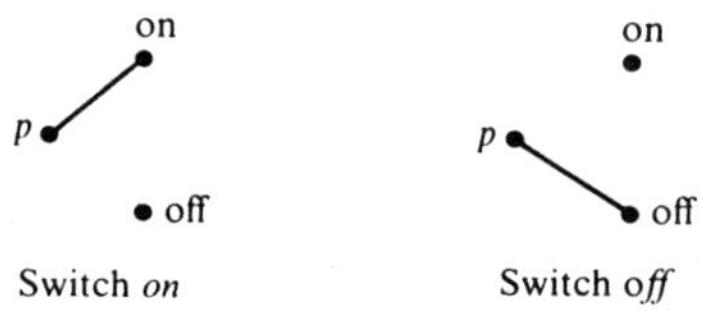

We call one state "on" and other state "off"; the switch will be either *on* or *off* and will never be in the middle. However, sometimes it is drawn in the middle to indicate that at any particular time it might be either on or off.

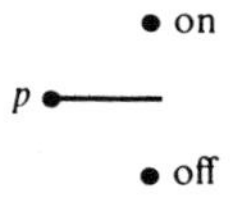

Suppose we connect a light to this switch so that the light is *on* when the switch is *on* and off when the switch is *off* (see Figure 3).

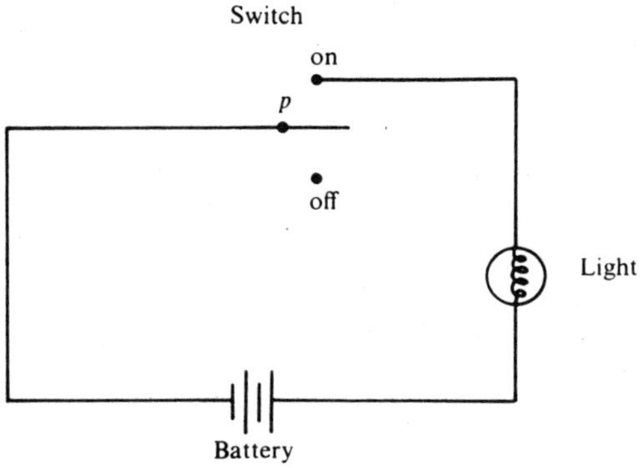

Figure 3 Schematic diagram showing how a light might be connected to a switch

*Optional chapter

A light is schematically drawn as . Since we need a power source, a battery, , is shown at the bottom of the figure. The light will light whenever the circuit is complete. Notice that the circuit is complete when p is switched *on* and not complete when p is *off.*

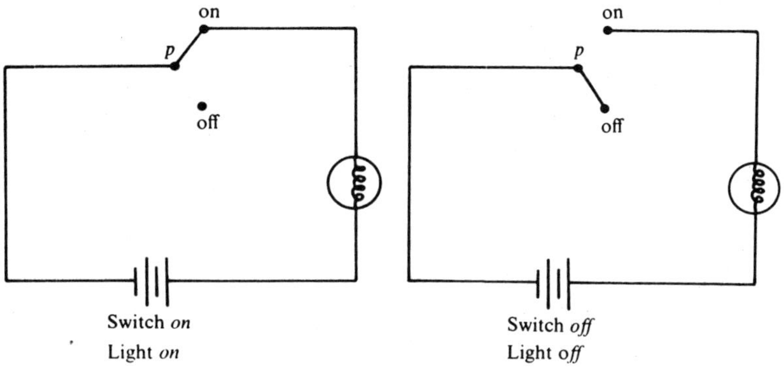

Figure 4 Schematic diagram showing the two possible states for Figure 3 *

Since these switches can assume either one of two states, *on* or *off,* and since our logical statements can assume either one of two states, *true* or *false,* it seems natural to define a relationship between a *switch* and a *logical statement.* Electronic circuits, which we will call *logical gates,* are devices with one or more inputs and one output. We will consider three of these:

1. **negation circuit** or *NOT-gate*
2. **conjunction circuit** or *AND-gate*
3. **disjunction circuit** or *OR-gate*

These circuits can be used to illustrate some of the logical principles we have been discussing. First, however, we must make the following agreements. We have been considering switches that are "on" or "off" and a light that is "on" or "off." Let us call switch p a logical proposition that is either true or false. If p is *true,* switch p will be represented as *on.* If p is *false,* switch p will be *off.* Similarly, we will let the light be the truth value of the simple or compound statements. Now let's consider some logical circuits.

10.2 Negation

Recall the truth table for negation.

p	$\sim p$
T	F
F	T

*Figure 4 shows two cases. However, we usually draw a switch as shown in Figure 3 to indicate either of the cases shown in Figure 4.

We see that the truth value for $\sim p$ is the opposite of the truth value for p. In terms of circuits, this means that, when switch p is "on," the light must be "off," and, if p is "off," the light must be "on." (You probably have some "negation switches" in your home.) The electrical circuit for negation is shown in Figure 5.

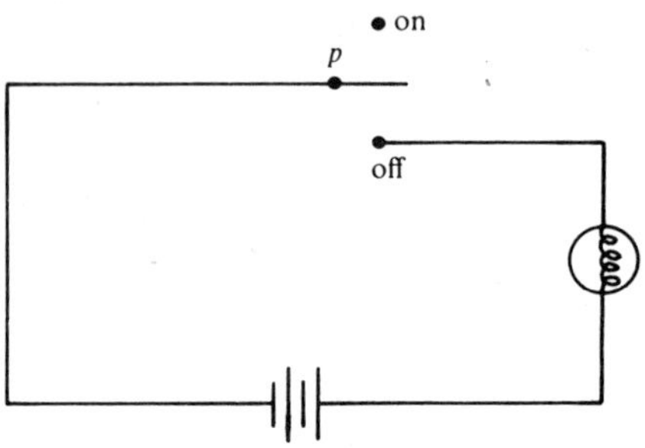

Figure 5 Negation circuit

10.3 Conjunction

Conjunction is a little more difficult, since two switches, p and q, are involved. The truth value of the light should indicate that it is true if and only if p and q are both true, as shown in the truth table for conjunction.

p	q	$p \wedge q$
T	T	T
T	F	F
F	T	F
F	F	F

Switches p and q must be connected *in series,* so that the circuit is complete if and only if both p and q are on (like Christmas lights that all go out if one goes out). This is shown in Figure 6.

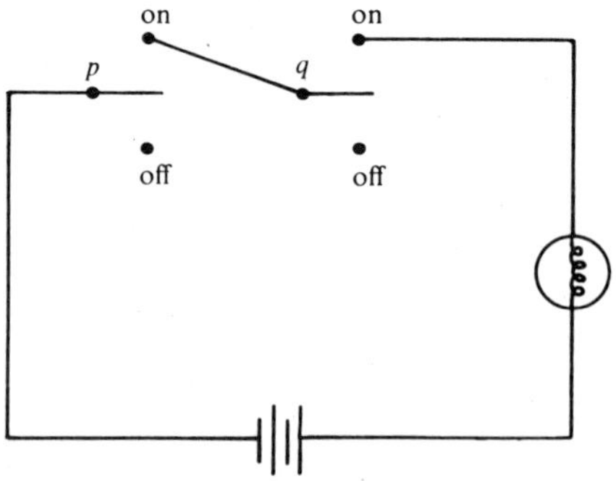

Figure 6 Conjunction circuit

10.4 Disjunction

We begin by studying the truth table for disjunction.

p	q	$p \vee q$
T	T	T
T	F	T
F	T	T
F	F	F

The truth value of $p \vee q$ is such that the light is *on* whenever either p or q is on. That is, p and q must be connected *in parallel,* so that the circuit is complete if either p or q is on (like Christmas lights that stay lit even if one light burns out). Study Figure 7, and then try to predict when the light will be *on.*

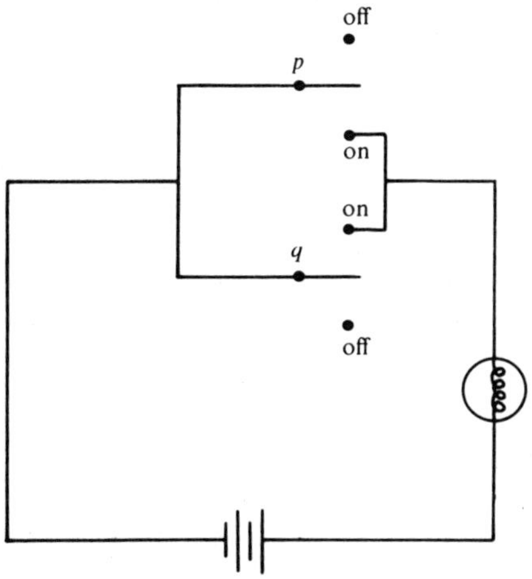

Figure 7 Disjunction circuit

10.5 Logical Gates

It is often necessary to combine the circuits of Figures 5, 6, and 7 into more complicated circuits that can be used to represent more complicated compound statements. When we do this, the schematics can become a little unmanageable. Therefore we will invent new notational symbols to stand for entire circuits. The first of these is called the *NOT-gate.* Instead of drawing the circuit shown in Figure 5, we will simply represent the entire circuit as follows:

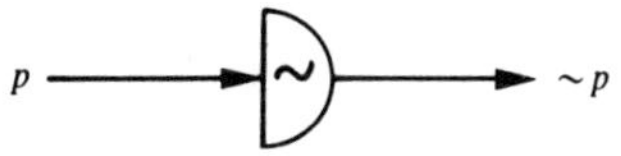

NOT-gate

Notice that the NOT-gate has a single input (proposition p) and a single output (proposition $\sim p$). The input p stands for the p-switch in Figure 5. The output $\sim p$ stands for the light in Figure 5. That is, current will flow out of the gate in those cases in which the light of Figure 5 is on and will not flow out of the gate in those cases in which the light is off.

The *AND-gate* (shown at the top of the next page), on the other hand, has two switches, p and q. These are symbolized as two inputs p and q. The single output $p \wedge q$ stands for the light in Figure

6. The following AND-gate can be used as a symbolic representation of the circuit shown in Figure 6:

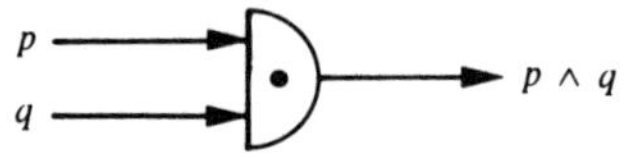

AND-gate

The *OR-gate* is similar, and the following diagram can be used to represent the circuit shown in Figure 7:

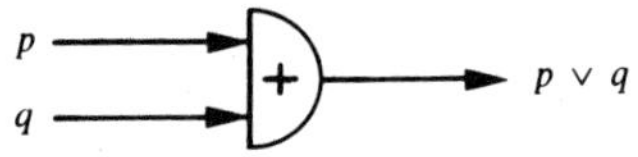

OR-gate

Logical circuits can be constructed to simulate the truth tables discussed in Chapter 2. For example, suppose we wish to find the truth values for $\sim(p \vee q)$. We should design the following circuit (remember that parentheses indicate the operation to be performed first):

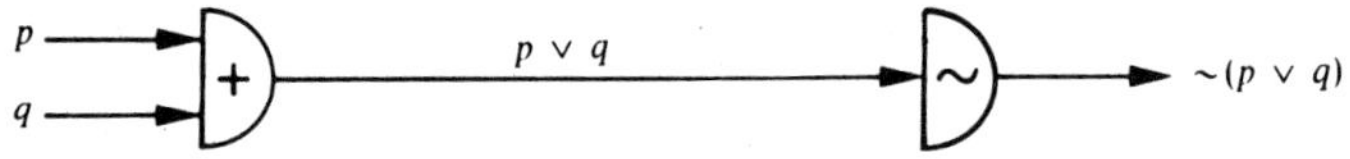

EXAMPLE 1 Design a circuit for $\sim p \wedge (\sim q)$.

Solution

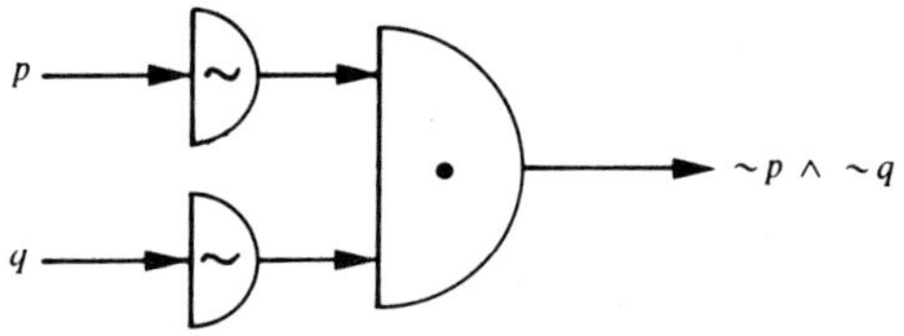

□

EXAMPLE 2 Design a circuit that will find the truth values for $(p \vee q) \wedge q$.

Solution We use gates to symbolize $(p \vee q) \wedge q$:

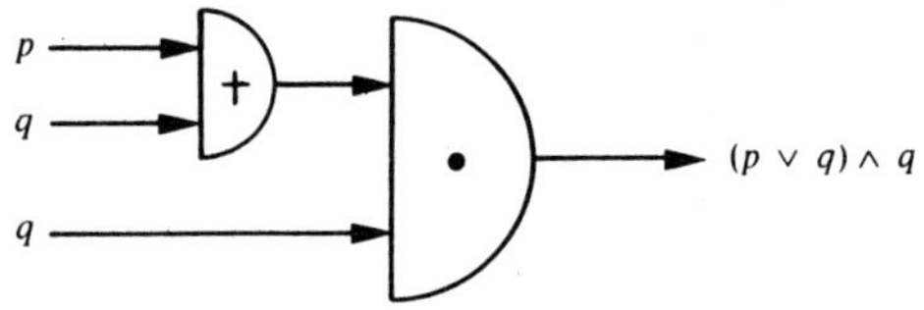

Next we must have some way of determining when the output values of $(p \vee q) \wedge q$ are true and when they are false. We can do this by again connecting a light to the circuit. When $(p \vee q) \wedge q$ is true, the light should be on; when it is false, the light should be off. We represent this connection as follows:

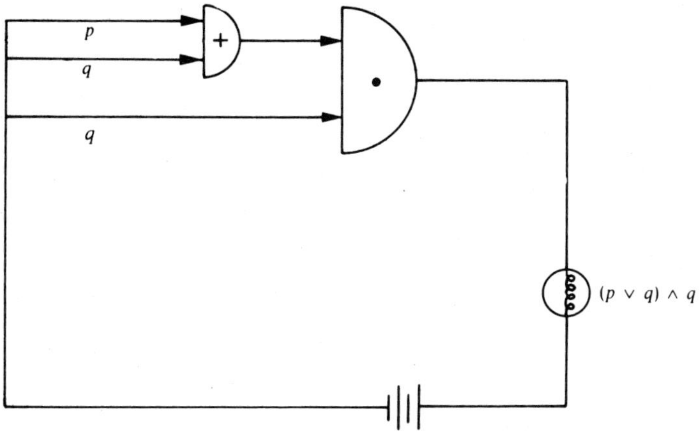

EXAMPLE 3 Design a circuit that will find the truth values of

$$\sim (p \vee q) \wedge [(p \vee q) \wedge q]$$

Solution

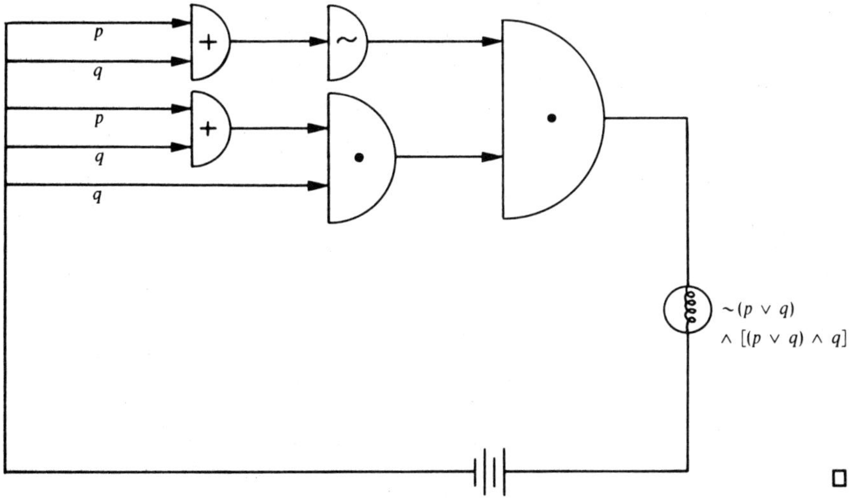

□

10.6 PROBLEM SET 10

1. Consider Figure 3. In your own words, explain when the light will be on and when it will be off.

2. What do the following circuit symbols mean?

a. 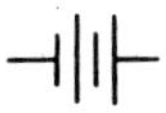

b. 

c. 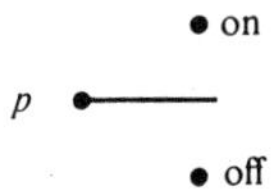

d. 

e. 

Using switches, design a circuit that would find the truth values for the statements in Problems 3–5 (answers are not unique).

3. $p \wedge q$ **4.** $p \wedge (\sim q)$ **5.** $\sim (p \vee q)$

6. Design a circuit for $p \rightarrow q$, using two switches.

7. Recall from Chapter 3 that $p \rightarrow q$ is defined to be $\sim [p \wedge (\sim q)]$. Design a circuit for $p \rightarrow q$ by using AND-gates and NOT-gates.

8. The cost of the circuit is often a factor in work with switching circuits. The circuit you constructed in Problem 7 by using the definition of the conditional, required three gates. Construct a truth table for $p \rightarrow q$ and $\sim p \vee q$. What do you notice about the truth values of the result? Design circuits for $p \rightarrow q$ by using only one OR-gate and one NOT-gate.

Using AND-gates, OR-gates, and NOT-gates, design circuits that would find the truth values for the sentences in Problems 9–12.

9. $p \wedge (q \vee r)$

10. $(p \wedge q) \vee (p \wedge r)$

11. $(\sim p \wedge q) \vee (p \vee \sim q)$

12. $\sim [(p \wedge q) \vee r]$

13. **a.** Suppose an engineer designs the following circuit:

$$\sim \{\sim [(p \vee q) \wedge (\sim p)] \wedge (\sim p)\}.$$

What will the circuit look like?

b. If each gate costs 2¢ and a company is going to manufacture one million items using this circuit, how much will the circuits cost?

c. Use truth tables to find the following.

i. $\sim \{\sim [(p \vee q) \wedge \sim p] \wedge (\sim p)\}$

ii. $p \vee q$

iii. What do you notice about the final column of each of these truth tables?

d. Draw a simpler circuit that will output the same values as those in part a.

e. If the company is going to manufacture one million items using this simpler circuit, how much will the circuits cost? How much will the company save by using this simpler circuit rather than the more complicated one?

Design the circuits in Problems 14–21.

14. $p \rightarrow \sim q$

15. $p \wedge \sim q$

16. $q \rightarrow p$

17. $\sim (\sim p \vee q)$

18. $\sim p \vee q$

19. $\sim p \rightarrow q$

20. $q \rightarrow \sim p$

21. $\sim p \wedge (\sim q)$

22. Alfie, Bogie, and Clyde are the members of a senate committee. Design a circuit that will output *yea* or *nay* depending on the way the majority of the committee members voted.

23. Suppose that the senate committee of Problem 22 has five members. Design a circuit that will output the result of the majority of the committee's vote.

*CHAPTER 11

ANALYSIS OF STATEMENTS

11.1 Mathematical Systems

In this book we've been considering propositions and statements and some of their relationships. We have also seen how logic can be applied to the theory of switching circuits and logical gates. We now wish to examine the underlying structure of these relationships.

We are all familiar with the algebraic structure of ordinary high-school algebra. Some of you may have studied the notions of fields or groups, which are examples of mathematical systems. The set of all possible propositions also gives rise to a mathematical system.

MATHEMATICAL SYSTEM

> **A mathematical system** is a nonempty set of elements along with one or more defined operations and one or more relations satisfying a specified set of axioms.

For symbolic logic we consider the following mathematical system.

Set: All propositions, which we denote by p, q, r, s, ...

Operations: The binary operations *or* ($\vee$) and *and* ($\wedge$).

Relation: Equivalence ($\leftrightarrow$); recall that equivalence between propositions means that they have the same truth values.

Axioms: The specified set of axioms is discussed in Sections 11.2–11.7.

11.2 Closure Property

The closure property says that a set is closed if the result of a particular binary operation is a member of the given set. In this chapter we are considering the set of all propositions and the operations of disjunction ($\vee$) and conjunction ($\wedge$). It is obvious that $p \wedge q$ and $p \vee q$ are propositions, so the closure property is satisfied.

*Optional chapter

11.2 Commutative Property

The commutative property for the operations of conjunction would say that $p \wedge q$ is the same as $q \wedge p$. We symbolize this as

$$(p \wedge q) \leftrightarrow (q \wedge p)$$

To *prove* the commutative property for conjunction, we need to construct a truth table (Table 15). Since the last column is all Ts, we see that $(p \wedge q) \leftrightarrow (q \wedge p)$ is a tautology, and the result has been proved.

Table 15
PROOF OF THE COMMUTATIVE PROPERTY FOR CONJUNCTION

p	q	$p \wedge q$	$q \wedge p$	$(p \wedge q) \leftrightarrow (q \wedge p)$
T	T	T	T	T
T	F	F	F	T
F	T	F	F	T
F	F	F	F	T

The commutative property for disjunction is

$$(p \vee q) \leftrightarrow (q \vee p),$$

and the proof of this property is left as a problem.

11.3 Associative Property

The associative property for conjunction is

$$[(p \wedge q) \wedge r] \leftrightarrow [p \wedge (q \wedge r)]$$

This can be shown to be a tautology, as we see in Table 16. The proof of the associative property for disjunction is left as a problem.

Table 16
PROOF OF THE ASSOCIATIVE PROPERTY FOR CONJUNCTION

p	q	r	$p \wedge q$	$(p \wedge q) \wedge r$	$q \wedge r$	$p \wedge (q \wedge r)$	$[(p \wedge q) \wedge r] \leftrightarrow [p \wedge (q \wedge r)]$
T	T	T	T	T	T	T	T
T	T	F	T	F	F	F	T
T	F	T	F	F	F	F	T
T	F	F	F	F	F	F	T
F	T	T	F	F	T	F	T
F	T	F	F	F	F	F	T
F	F	T	F	F	F	F	T
F	F	F	F	F	F	F	T

11.4 Distributive Property

There are many possible distributive properties that we might check. Let us begin by using truth tables to show that conjunction is distributive over disjunction:

$$[p \wedge (q \vee r)] \leftrightarrow [(p \wedge q) \vee (p \wedge r)]$$

This looks complicated, but if we take one step at a time, the truth table is not too difficult, as is shown in Table 17.

Table 17
PROOF OF THE DISTRIBUTIVE PROPERTY – CONJUNCTION OVER DISJUNCTION

$p\ q\ r$	$q \vee r$	$p \wedge (q \vee r)$	$p \wedge q$	$p \wedge r$	$(p \wedge q) \vee (p \wedge r)$	$[p \wedge (q \vee r)] \leftrightarrow [(p \wedge q) \vee (p \wedge r)]$
T T T	T	T	T	T	T	T
T T F	T	T	T	F	T	T
T F T	T	T	F	T	T	T
T F F	F	F	F	F	F	T
F T T	T	F	F	F	F	T
F T F	T	F	F	F	F	T
F F T	T	F	F	F	F	T
F F F	F	F	F	F	F	T

Another distributive property that holds is for disjunction over conjunction:

$$[p \vee (q \wedge r)] \leftrightarrow [(p \vee q) \wedge (p \vee r)]$$

The proof of this property is left as a problem.

11.5 Identity Elements

Let's use the symbol "1" to represent a proposition that is permanently true (a tautology). Similarly, "0" will represent a proposition that is permanently false (a contradiction). For example, we say $[p \wedge (\sim p)] \Leftrightarrow 0$ and $[p \vee (\sim p)] \Leftrightarrow 1$.

We have therefore found an identity element for conjunction, since $(p \wedge 1) \Leftrightarrow p$ for any proposition p. That is, 1 is an identity element for conjunction. Similarly, it can be shown that 0 is an identity for disjunction. We could have talked about identities without introducing the 1 and 0 symbolism by saying

p *and* (a tautology) $\Leftrightarrow p$
p *or* (a contradiction) $\Leftrightarrow p$

11.6 Inverse and Property of Complementation

For every proposition p, there exists a proposition $\sim p$ so that

$$[p \wedge (\sim p)] \Leftrightarrow 0 \quad \text{and} \quad [p \vee (\sim p)] \Leftrightarrow 1$$

Does this mean we have the inverse property in the set? Notice that, for the inverse property, we would have liked to obtain

$$[p \wedge (\sim p)] \Leftrightarrow 1 \quad \text{and} \quad [p \vee (\sim p)] \Leftrightarrow 0$$

but these are *false*. We must say that the set does not have the inverse property, but we do see a pattern. The operations $\wedge$ and $\vee$ are more closely related than some of the other operations we have investigated. Let us call this relationship the **property of complementation.**

PROPERTY OF COMPLEMENTATION

$(p \wedge \sim p) \Leftrightarrow 0$
$(p \vee \sim p) \Leftrightarrow 1$

11.7 Boolean Algebra

The properties of this mathematical system for symbolic logic can be generalized. Instead of applying them solely to symbolic logic, we can also apply them to the algebra of switching circuits and to the algebra of sets. To do this, we will define an abstract mathematical system called a *Boolean algebra,* named after George Boole (1815–1864). As we pointed out in Chapter 1, this idea had occurred to others, but Boole was the first to produce a workable system, in his book *The Mathematical Analysis of Logic.* By and large, neither the philosophers nor the mathematicians of his century showed much interest in this invention. One of the reasons for their lack of interest was the clumsy notation Boole chose for his algebra. Today the notation is greatly improved, and this abstract algebra of Boole's is regarded as a mathematical system in which no meanings whatever are assigned to the symbols. Herein lies much of its strength, since many different interpretations can be given to the Boolean symbols. It was this algebra that opened the door for symbolic logic. Today Boolean algebra goes beyond the applications of symbolic logic and is an important discipline in its own right.

To define a Boolean algebra, we need to use symbols that can be applied not only to logic but to other applications as well.

BOOLEAN ALGEBRA

Let S be a set of elements $a, b, c, \ldots$ with two binary operations $\otimes$ and $\oplus$ and an equivalence relation $\equiv$. Then S is a **Boolean algebra** if the following axioms are satisfied:

1. **Closure:** If a, b are elements of S, then

   $(a \oplus b)$ and $(a \otimes b)$ are elements of S.

2. **Commutative:** If a, b are elements of S, then

   $a \oplus b \equiv b \oplus a$ and $a \otimes b \equiv b \otimes a$.

3. **Associative:** If a, b, c are elements of S, then

   $(a \oplus b) \oplus c \equiv a \oplus (b \oplus c)$ and $(a \otimes b) \otimes c \equiv a \otimes (b \otimes c)$.

4. **Distributive:** If a, b, c are elements of S, then

   $a \oplus (b \otimes c) \equiv (a \oplus b) \otimes (a \oplus c)$ and $a \otimes (b \oplus c) \equiv (a \otimes b) \oplus (a \otimes c)$.

5. **Identity:** There exist elements e and z in S such that, for each a of S,

   $a \oplus z \equiv a$ and $a \otimes e \equiv a$.

6. **Complementation:** For each a in S there exists an element a' in S such that

   $a \oplus a' \equiv e$ and $a \otimes a' \equiv z$.

11.8 Boolean Algebra of Logic, Circuits, and Sets

As you can see, our definition of a Boolean algebra is quite abstract. However, if we assign certain meanings to the symbols, their relationship to logic, circuits, and sets becomes clear.

Relationship to Logic:
The set S is the set of propositions;
$\oplus$ is the operation disjunction ($\vee$);
$\otimes$ is the operation conjunction ($\wedge$);
$\equiv$ is the equivalence relation ($\leftrightarrow$);
e is a tautology (1);
z is a contradiction (0);
a' is the negation of a proposition a ($\sim a$).

Relationship to Switching Circuits:
The set S is the set of possible switching circuits states, 0 and 1 (representing *off* and *on*, respectively);
$\oplus$ is the operation of connecting switches in parallel (+);
$\otimes$ is the operation of connecting switches in series ($\cdot$);
$\equiv$ is the relation = (which says that two circuits have the same output value, 0 or 1);
e is a switching circuit that is always on (1);
z is a switching circuit that is always off (0);
a' is the negation switch ($0' = 1$ and $1' = 0$).

Relationship to Sets:
The set S is the set of all subsets of a universal set;
$\oplus$ is the operation of union ($\cup$);
$\otimes$ is the operation of intersection ($\cap$);
$\equiv$ is the relation of equality (=);
e is the universal set (U);
z is the empty set ($\emptyset$);
a' is the complement of a set A ($\overline{A}$).

The laws of a Boolean algebra, along with their interpretations for set algebra and symbolic logic, are summarized in Table 18.

Table 18
RELATIONSHIP AMONG BOOLEAN ALGEBRA, SET ALGEBRA, AND SYMBOLIC LOGIC

BOOLEAN ALGEBRA	SET ALGEBRA	SYMBOLIC LOGIC
COMMUTATIVE LAWS:		
1. $a \oplus b \equiv b \oplus a$	$A \cup B = B \cup A$	$(p \vee q) \Leftrightarrow (q \vee p)$
2. $a \otimes b \equiv b \otimes a$	$A \cap B = B \cap A$	$(p \wedge q) \Leftrightarrow (q \wedge p)$
ASSOCIATIVE LAWS:		
3. $(a \oplus b) \oplus c \equiv a \oplus (b \oplus c)$	$(A \cup B) \cup C = A \cup (B \cup C)$	$[(p \vee q) \vee r] \Leftrightarrow [p \vee (q \vee r)]$
4. $(a \otimes b) \otimes c \equiv a \otimes (b \otimes c)$	$(A \cap B) \cap C = A \cap (B \cap C)$	$[(p \wedge q) \wedge r] \Leftrightarrow [p \wedge (q \wedge r)]$
DISTRIBUTIVE LAWS:		
5. $a \otimes (b \oplus c) \equiv (a \otimes b) \oplus (a \otimes c)$	$A \cap (B \cup C) = (A \cap B) \cup (A \cap C)$	$[p \wedge (q \vee r)] \Leftrightarrow [(p \wedge q) \vee (p \wedge r)]$
6. $a \oplus (b \otimes c) \equiv (a \oplus b) \otimes (a \oplus c)$	$A \cup (B \cap C) = (A \cup B) \cap (A \cup C)$	$[p \vee (q \wedge r)] \Leftrightarrow [(p \vee q) \wedge (p \vee r)]$

BOOLEAN ALGEBRA	SET ALGEBRA	SYMBOLIC LOGIC
IDEMPOTENT LAWS:		
7. $a \oplus a \equiv a$	$A \cup A = A$	$(p \vee p) \Leftrightarrow p$
8. $a \otimes a \equiv a$	$A \cap A = A$	$(p \wedge p) \Leftrightarrow p$
IDENTITY LAWS:		
9. $a \oplus e \equiv e$	$A \cup U = U$	$(p \vee 1) \Leftrightarrow 1$
10. $a \oplus z \equiv a$	$A \cup \emptyset = A$	$(p \vee 0) \Leftrightarrow p$
11. $a \otimes e \equiv a$	$A \cap U = A$	$(p \wedge 1) \Leftrightarrow p$
12. $a \otimes z \equiv z$	$A \cap \emptyset = \emptyset$	$(p \wedge 0) \Leftrightarrow 0$
COMPLEMENT LAWS:		
13. $a \oplus a' \equiv e$	$A \cup \overline{A} = U$	$[p \vee (\sim p)] \Leftrightarrow 1$
14. $a \otimes a' \equiv z$	$A \cap \overline{A} = \emptyset$	$[p \wedge (\sim p)] \Leftrightarrow 0$
15. $e' \equiv z$	$\overline{U} = \emptyset$	$-1 \Leftrightarrow 0$
16. $z' \equiv e$	$\overline{\emptyset} = U$	$-0 \Leftrightarrow 1$
17. $(a')' \equiv a$	$\overline{\overline{A}} = A$	$\sim(\sim p) \Leftrightarrow p$

Table 18 Laws of Boolean Algebra

11.9 PROBLEM SET 11

In the following problems, let S be the set for all propositions.

1. For the set S, prove the commutative property for disjunction.

2. In S, prove the associative property for disjunction.

3. In S, is disjunction distributive over conjunction? Support your answer.

4. In S, is negation distributive over conjunction? Support your answer.

5. What do we mean by an "identity element"?

6. What is a Boolean algebra?

7. Does the set S with the operations of conjunction and disjunction form a Boolean algebra?

CHAPTER 12

SUMMARY AND REVIEW

12.1 Outline

I. Simple and Compound Statements
   A. *Definition:* A **statement** is a sentence that is either true or false, but not both true and false.
      1. *Simple statement* — no connective
      2. *Compound statement* — one or more connectives
   B. Common operators
      1. **Conjunction** — "and"; $p \wedge q$.
      2. **Disjunction** — "or"; $p \vee q$.
      3. **Negation** — "not"; $\sim p$
      4. **Implication** — "if ... then"; $p \rightarrow q$.
      5. **Equivalence** — "if and only if"; $p \leftrightarrow q$.
   C. Additional connectives
      1. Either p or q: $(p \vee q) \wedge [\sim(p \wedge q)]$
      2. Neither p nor q: $\sim p \wedge (\sim q)$
      3. p unless q: $\sim q \rightarrow p$
      4. p because q: $q \rightarrow p$
      5. No p is q: $p \rightarrow (\sim q)$

II. Truth Tables

III. Summary of Important Operators

p	q	$\sim p$	$p \wedge q$	$p \vee q$	$p \rightarrow q$	$p \leftrightarrow q$
T	T	F	T	T	T	T
T	F	F	F	T	F	F
F	T	T	F	T	T	F
F	F	T	F	F	T	T

IV. The Conditional
   A. If–then: $p \rightarrow q$
   B. Converse: $q \rightarrow p$
   C. Inverse: $\sim p \rightarrow (\sim q)$
   D. Contrapositive: $\sim q \rightarrow (\sim p)$
   E. Equivalent statements
      1. The implication and contrapositive are equivalent.
      2. The converse and inverse are equivalent.

V. Fundamental Tautologies
   A. *Definition:* A **tautology** is a statement that is true for every case of its truth table.
   B. Implication ($\Rightarrow$)

- C. Logical equivalence ($\Leftrightarrow$)
- D. Trivial tautology: $p \Leftrightarrow p$
- E. Law of double negation: $p \Leftrightarrow \sim(\sim p)$
- F. Law of the excluded middle: $p \vee (\sim p)$
- G. Direct reasoning: $$\begin{array}{l} p \to q \\ \underline{p \qquad} \\ \therefore\ q \end{array}$$
- H. Indirect reasoning: $$\begin{array}{l} p \to q \\ \underline{\sim q \qquad} \\ \therefore\ \sim p \end{array}$$
- I. Law of transitivity: $$\begin{array}{l} p \to q \\ \underline{q \to r} \\ \therefore\ p \to r \end{array}$$
- J. Law of contraposition: $(p \to q) \Leftrightarrow [\sim q \to (\sim p)]$
- K. Disjunctive syllogism: $$\begin{array}{l} p \vee q \\ \underline{\sim p \qquad} \\ \therefore\ q \end{array}$$

VI. Logical Fallacies
- A. Assuming the antecedent
- B. Denying the consequent
- C. False chain patterns

VII. Proof of Logical Arguments
- A. Truth and validity
- B. Quantifiers
  1. Universal
  2. Existential
- C. Truth values of quantified sentences
  1. A statement with a universal quantifier is true if and only if the replacement set of the variable is the same as the universal set for the problem.
  2. A statement with an existential quantifier is true if and only if the replacement set of the variable is nonempty.
- D. Steps in analyzing logical arguments
  1. Translate the English into symbols.
  2. Simplify the symbolic argument.
  3. Translate the conclusion back into English.
- E. Using Euler circles to prove logical arguments

VIII. Logic Circuits
- A. Basic switching circuits
- B. Logical gates

IX. Analysis of Statements
- A. A *mathematical system* is a nonempty set of elements along with one or more defined operations and one or more relations satisfying a specified set of axioms.
- B. Properties
  1. Closure
  2. Associative

3. Commutative
4. Identity
5. Complementation

C. Boolean algebra

1. *Definition:* A **Boolean algebra** is a mathematical system with two operations satisfying the closure, commutative, associative, distributive, identity, and complementation properties for both operations.
2. Relationship to symbolic logic
3. Relationship to switching circuits
4. Relationship to sets

12.2 PROBLEM SET 12

1. Discuss the difference between definitions and theorems.

2. Complete the following truth table.

p	q	$\sim p$	$p \wedge q$	$p \vee q$	$p \rightarrow q$	$p \leftrightarrow q$

3. Construct a truth table for $\sim (p \wedge q)$

4. Design a logical circuit for $\sim (p \wedge q)$

5. Show that $\sim (p \wedge q) \leftrightarrow [(\sim p) \vee (\sim q)]$

6. Construct a truth table for $\sim [(\sim p \vee q) \wedge p]$

7. Using AND-gates, OR-gates, and NOT-gates, design a circuit for the statement given in Problem 6.

8. State and prove the principle of direct reasoning.

9. Give an example of indirect reasoning.

10. Give an example of a common fallacy of logic, and show why it is a fallacy.

11. Let p: P is a prime number; q: $P + 2$ is a prime number. Translate the following statements into symbolic form.

a. $p \rightarrow q$ **b.** $(p \vee q) \wedge [\sim (p \wedge q)]$

12. Consider the statement "All computers are incapable of self-direction."

a. Translate this statement into symbolic form.
b. Write the contrapositive of the statement.

13. Consider the statement "If Rosamond doesn't go to college, then she is not liberated." Write the converse, inverse, and contrapositive of the statement.

14. Find the truth values of each of the following statements.

a. $7 \cdot 9 = 63$ if and only if $5 + 10 = 12$

b. $6 + 2 = 10$ if and only if today is February 30.

15. Is the following valid?

$$\begin{array}{l} p \rightarrow q \\ \sim q \\ \hline \therefore \sim p \end{array}$$

Can you support your answer?

16. Test the validity of the following argument. Support your answer.

1. If Andrew attends the lecture, we can use his notes.
2. Andrew does not attend the lecture.

Therefore we cannot use his notes.

17. Form a valid conclusion using all the statements.

1. If I attend to my duties, I am rewarded.
2. If I am lazy, I am not rewarded.
3. I am lazy.

18. Form a valid conclusion using all the statements.

1. If Jack and Jill go up the hill, then they are fetching a pail of water.
2. If Jack and Jill are fetching a pail of water, then Jack will certainly tumble.
3. If Jack falls, then Jill will also fall.
4. If Jill falls, then it will cost Jack a lot of money.

19. Show that disjunction is associative.

20. Write each of the following open sentences as a true statement.

a. $5x + 7x = 12x$

b. $15y < 10$

c. $z + 2 = 5$

d. $x + y = 7$

e. $a + b = b + a$

21. **a.** What is a Boolean algebra?

b. Why is it useful?

22. Use Euler circles to show: Some dogs have fleas.

23. **a.** Is the quantifier used in Problem 22 universal or existential?

b. Write an appropriate symbol for the quantifier used in Problem 22.

24. In your own words, discuss universal and existential quantifiers, and give examples of each.

25. Use Euler circles to test the validity of the following argument.

1. All dancers are eccentric.
2. All eccentric people are rich.
3. No boring people are nondancers.
4. All eccentric people are singers.

Therefore no singers are rich.

26. Can any other valid conclusions be forced from the premises in Problem 25?

In Problems 27–31, test the validity of the given arguments. Support your answers.

27. If Ted works hard, then he succeeds.
Ted succeeds.
Therefore Ted works hard.

28. If your tire tread does not cover the head of Lincoln (on a penny), then your tire is worn thin.
Your tire tread is worn thin.
Therefore your tire tread does not cover the head of Lincoln.

29. If your tire tread does not cover the head of Lincoln (on a penny), then your tire is worn thin.
Your tire tread does not cover the head of Lincoln.
Therefore your tire is worn thin.

30. If you want peace, prepare for war.
You do not prepare for war.
Therefore you do not want peace.

31. If your parents both have type O blood, then you have type O blood.
You have type O blood.
Therefore your parents both have type O blood.

32. Which of the following statements are true?
a. If $111 + 1 = 1000$, then I'm a monkey's uncle.
b. If $6 + 2 = 10$ and $7 + 2 = 9$, then $5 + 2 = 7$.
c. If $6 + 2 = 10$ or $5 + 2 = 7$, then $7 + 2 = 9$.
d. If $5 + 2 = 7$ and $7 + 2 \neq 9$, then $6 + 2 = 10$.
e. If the moon is made of green cheese, then $5 + 7 = 57$.

33. Write the following pair of statements as one statement by using the appropriate connective.
i. If $ad = bc$, then $a/b = c/d$.
ii. If $a/b = c/d$, then $ad = bc$.

APPENDIX ANSWERS

Problem Set 1

1. *a*, *b*, *c*, and *d* are statements. **2.** *a*, *b*, *c*, and *e* are statements. **3.** *a*, *b*, *c*, and *e* are statements. **4.** *a*, *c*, and *d* are statements. **5.** See Table 1. **6.** See Table 2. **7.** See Table 3. **8. a.** Some mathematicians are not ogres. **b.** No integer is negative. **c.** All people pay taxes. **d.** Some even integers are divisible by 5. **9. a.** Some counting numbers are not divisible by 1. **b.** No apples are rotten. **c.** All integers are odd. **d.** Some triangles are squares. **10. a.** Some women are not intelligent. **b.** No rectangles are squares. **c.** All rectangles are squares. **d.** Some nice people are dangerous. **11.** $p \wedge q$ **12.** $\sim(p \wedge q)$ **13.** $p \vee (\sim q)$ **14.** $p \wedge (\sim q)$ **15.** $(p \vee q) \wedge [\sim(p \wedge q)]$ **16.** Paul is peculiar and likes to read math textbooks. **17.** Paul is not peculiar and likes to read math textbooks. **18.** It is not the case that Paul is peculiar and likes to read math textbooks. **19.** Paul is peculiar or does not like to read math textbooks. **20.** Paul is not peculiar or does not like to read math textbooks. **21.** Let *e*: Pat is eating. *d*: Pat is drinking. *t*: Pat is having a good time. Then $e \wedge d \wedge t$. **22.** Let *a*: Sam will accept the nomination. Then $\sim a$. **23.** Let *n*: You will go out tonight. *m*: You will go out tomorrow. Then $\sim n \wedge (\sim m)$. **24.** Let ℓ: Fat Albert lives to eat. *k*: Fat Albert eats to live. Then $\ell \wedge (\sim k)$. **25.** Let *j*: The decision depends on judgment. *i*: The decision depends on intuition. *p*: The decision depends on who paid the most. Then $(j \vee i) \wedge (\sim p)$. **26.** F; for all nonzero real numbers x, $|x| > 0$ **27.** F; for all nonnegative real numbers x, $\sqrt{x^2} = x$ **28.** T **29.** T (it is actually true for all real numbers) **30.** F; there exists some x such that $|x + 4| < 0$ **31.** T

Problem Set 2

1.

p	q	$\sim p$	$\sim q$	$\sim p \wedge \sim q$
T	T	F	F	F
T	F	F	T	F
F	T	T	F	F
F	F	T	T	T

2.

p	q	$p \wedge q$	$\sim(p \wedge q)$
T	T	T	F
T	F	F	T
F	T	F	T
F	F	F	T

3.

p	q	$\sim p$	$\sim q$	$\sim p \vee \sim q$
T	T	F	F	F
T	F	F	T	T
F	T	T	F	T
F	F	T	T	T

4.

p	q	$p \wedge q$	$p \wedge (p \vee q)$
T	T	T	T
T	F	T	T
F	T	T	F
F	F	F	F

5.

r	s	$\sim s$	$r \wedge (\sim s)$
T	T	F	F
T	F	T	T
F	T	F	F
F	F	F	F

6.

s	$\sim s$	$s \vee (\sim s)$
T	F	T
F	T	T

7.

r	s	$r \wedge s$	$\sim s$	$(r \wedge s) \vee (\sim s)$
T	T	T	F	T
T	F	F	T	T
F	T	F	F	F
F	F	F	T	T

8.

r	s	$\sim r$	$\sim s$	$\sim r \vee (\sim s)$
T	T	F	F	F
T	F	F	T	T
F	T	T	F	T
F	F	T	T	T

9.

p	q	$p \vee q$	$p \wedge q$	$(p \vee q) \vee (p \wedge q)$
T	T	T	T	T
T	F	T	F	T
F	T	T	F	T
F	F	F	F	F

10.

p	q	$\sim p$	$\sim q$	$\sim p \wedge q$	$(\sim p \wedge q) \vee (\sim q)$
T	T	F	F	F	F
T	F	F	T	F	T
F	T	T	F	T	T
F	F	T	T	F	T

11.

p	q	$\sim q$	$p \wedge (\sim q)$	$[p \wedge (\sim q)] \wedge p$
T	T	F	F	F
T	F	T	T	T
F	T	F	F	F
F	F	T	F	F

12.

p	q	$\sim p$	$q \vee (\sim p)$	$p \wedge [q \vee (\sim p)]$	$\sim q$	$p \wedge (\sim q)$	$\{p \wedge [q \vee (\sim p)\} \vee [p \wedge (\sim q)]$
T	T	F	T	T	F	F	T
T	F	F	F	F	T	T	T
F	T	T	T	F	F	F	F
F	F	T	T	F	T	F	F

13.

p	q	r	$p \wedge q$	$(p \wedge q) \wedge r$
T	T	T	T	T
T	T	F	T	F
T	F	T	F	F
T	F	F	F	F
F	T	T	F	F
F	T	F	F	F
F	F	T	F	F
F	F	T	F	F

14.

p	q	r	$p \wedge q$	$\sim r$	$(p \wedge q) \wedge (\sim r)$
T	T	T	T	F	F
T	T	F	T	T	T
T	F	T	F	F	F
T	F	F	F	T	F
F	T	T	F	F	F
F	T	F	F	T	F
F	F	T	F	F	F
F	F	F	F	T	F

15.

p	q	r	$p \vee q$	$\sim r$	$(p \vee q) \wedge (\sim r)$	$[(p \vee q) \wedge (\sim r)] \wedge r$
T	T	T	T	F	F	F
T	T	F	T	T	T	F
T	F	T	T	F	F	F
T	F	F	T	T	T	F
F	T	T	T	F	F	F
F	T	F	T	T	T	F
F	F	T	F	F	F	F
F	F	F	F	T	F	F

16.

p	q	r	$\sim p$	$q \vee (\sim p)$	$p \wedge [q \vee (\sim p)]$	$\{p \wedge [q \vee (\sim p)]\} \vee r$
T	T	T	F	T	T	T
T	T	F	F	T	T	T
T	F	T	F	F	F	T
T	F	F	F	F	F	F
F	T	T	T	T	F	T
F	T	F	T	T	F	F
F	F	T	T	T	F	T
F	F	F	T	T	F	F

17. 2, 3, and 4 are true. **18.** (5) Mankind pollutes the environment, and mankind will not survive. True when r is T and s is F. (6) Mankind will survive, or mankind will not survive. Always true. (7) Mankind pollutes the environment and will survive, or mankind will not survive. True in all cases except when r is F and s is T. (8) Mankind does not pollute the environment, or mankind will not survive. True in all cases except when r and s are both true. **19.** yes **20.** yes **21.** yes **22.** yes **23.** **a.** 2 **b.** 4 **c.** 8 **24.** 16 **25.** 2^n

Problem Set 3

1. If you succeed, then you value others' judgments. Let s: You succeed; v: You value others' judgments. $s \rightarrow v$ **2.** If I get paid, then I will go on Saturday. Let p: I get paid; g: I will go on Saturday. $p \rightarrow g$ **3.** If there is time enough, then everything happens to everybody sooner or later. Let t: There is time enough; h: Everything happens to everybody sooner; i: Everything happens to everybody later. $t \rightarrow (h \vee i)$ **4.** If it is work, then it is noble. Let w: It is work; n: It is noble. $w \rightarrow n$ **5.** If $|x - a| < \delta$, then $|f(x) - L| < \varepsilon$. Let q: $|x - a| < \delta$; v: $|f(x) - L| < \epsilon$. $q \rightarrow v$ **6.** If combustion occurs, then oxygen is present. Let o: Oxygen is present; b: Combustion occurs. $b \rightarrow o$ **7.** If combustion occurs, then oxygen is present. Let o: Oxygen is present; b: Combustion occurs. $b \rightarrow o$ **8.** If John is reasonable, then he knows all the facts. Let j: John is reasonable; f: John knows all the facts. $j \rightarrow f$ **9.** If we make a proper use of those means which the God of Nature has placed in our power, then we are not weak. Let u: We make a proper use of those means which the God of Nature has placed in our power; w: We are weak. $u \rightarrow (\sim w)$ **10.** If x is non-negative, then $x < 5$. Let n: x is negative; m: $x < 5$. $\sim n \rightarrow m$ **11.** If $y = 2$, then $y^2 = 4$. Let e: $y = 2$; f: $y^2 = 4$. $e \rightarrow f$ **12.** If a polygon is a square, then it is a rectangle. Let s: A polygon is a square; r: A polygon is a rectangle. $s \rightarrow r$ **13.** If it is a triangle, then it is a polygon. Let t: It is a triangle. p: It is a rectangle. $t \rightarrow p$ **14.** If it is a prime greater than 2, then it is an odd prime. Let p: It is a prime greater than 2. d: It is an odd prime. $p \rightarrow d$ **15.** true **16.** true **17.** true **18.** true **19.** true **20.** If you like mathematics, then you like this text. **21.** You do not like mathematics, or you like this text. **22.** If you do not like this text, then you do not like mathematics. **23.** If you like or don't like mathematics, then you like this text. **24.** If you don't like mathematics, then you like this text.

25.

p	q	$\sim p$	$\sim p \rightarrow q$	$p \rightarrow (\sim p \rightarrow q)$
T	T	F	T	T
T	F	F	T	T
F	T	T	T	T
F	F	T	F	T

26.

p	q	$p \vee q$	$\sim q$	$p \wedge \sim q$	$(p \vee q) \vee (p \wedge \sim q)$
T	T	T	F	F	T
T	F	T	T	T	T
F	T	T	F	F	T
F	F	F	T	F	F

27.

p	q	$p \wedge q$	$\sim q$	$p \rightarrow (\sim q)$	$(p \wedge q) \wedge (p \rightarrow \sim q)$
T	T	T	F	F	F
T	F	F	T	T	F
F	T	F	F	T	F
F	F	F	T	T	F

28.

p	q	$p \wedge q$	$(p \wedge q) \rightarrow p$
T	T	T	T
T	F	F	T
F	T	F	T
F	F	F	T

29.

p	q	$p \to q$	$\sim q$	$q \to (\sim q)$	$(p \to p) \to [q \to (\sim q)]$
T	T	T	F	F	F
T	F	T	T	T	T
F	T	T	F	F	F
F	F	T	T	T	T

30.

p	q	$\sim p$	$p \wedge q$	$\sim (p \wedge q)$	$\sim p \to \sim (p \wedge q)$
T	T	F	T	F	T
T	F	F	F	T	T
F	T	T	F	T	T
F	F	T	F	T	T

31.

p	q	$p \to q$	$\sim q$	$\sim p$	$\sim q \to (\sim p)$	$(p \to q) \to [\sim q \to (\sim p)]$
T	T	T	F	F	T	T
T	F	F	T	F	F	T
F	T	T	F	T	T	T
F	F	T	T	T	T	T

32.

p	q	$p \wedge q$	$p \vee q$	$(p \wedge q) \to (p \vee q)$
T	T	T	T	T
T	F	F	T	T
F	T	F	T	T
F	F	F	F	T

33.

p	q	$p \to q$	$p \vee q$	$(p \vee q) \to q$	$(p \to q) \to [(p \vee q) \to q]$
T	T	T	T	T	T
T	F	F	T	F	T
F	T	T	T	T	T
F	F	T	F	T	T

34.

p	q	$p \to q$	$(p \to q) \to p$	$[(p \to q) \to p] \to (p \to q)$
T	T	T	T	T
T	F	F	T	F
F	T	T	F	T
F	F	T	F	T

35.

p	q	r	$p \to q$	$(p \to q) \to r$	$q \to p$	$(q \to p) \to r$	$[(p \to q) \to r] \to [(q \to p) \to r]$
T	T	T	T	T	T	T	T
T	T	F	T	F	T	F	T
T	F	T	F	T	T	T	T
T	F	F	F	T	T	F	F
F	T	T	T	T	F	T	T
F	T	F	T	F	F	T	T
F	F	T	T	T	T	T	T
F	F	F	T	F	T	F	T

36.

p	q	$\sim p$	$\sim p \wedge q$	$p \vee (\sim p \wedge q)$	$p \vee q$	$[p \vee (\sim p \wedge q)] \to (p \vee q)$
T	T	F	F	T	T	T
T	F	F	F	T	T	T
F	T	T	T	T	T	T
F	F	T	F	F	F	T

37.

p	q	$p \vee q$	$\sim p$	$\sim p \wedge q$	$p \vee (\sim p \wedge q)$	$(p \vee q) \rightarrow [p \vee (\sim p \wedge q)$
T	T	T	F	F	T	T
T	F	T	F	F	T	T
F	T	T	T	T	T	T
F	F	F	T	F	F	T

38. CONVERSE: If you go to jail, then you break the law. **INVERSE:** If you do not break the law, then you do not go to jail. **CONTRAPOSITIVE:** If you do not go to jail, then you do not break the law. **39. CONVERSE:** If q, then p. **INVERSE:** If $\sim p$, then $\sim q$. **CONTRAPOSITIVE:** If $\sim q$, then $\sim p$. **40.** CONVERSE: $\sim q \rightarrow \sim p$ **INVERSE:** $p \rightarrow q$ **CONTRAPOSITIVE:** $q \rightarrow p$ **41. CONVERSE:** If you are happy, then the sun shines. **INVERSE:** If the sun doesn't shine, then you are not happy. **CONTRAPOSITIVE:** If you are not happy, then the sun doesn't shine. **42. CONVERSE:** If you don't have friends, then your car is not air-conditioned. **INVERSE:** If your car is air-conditioned, then you have friends. **CONTRAPOSITIVE:** If you have friends, then your car is air-conditioned. **43. CONVERSE:** If I will go on Saturday, then I get paid. **INVERSE:** If I don't get paid, then I will not go on Saturday. **CONTRAPOSITIVE:** If I will not go on Saturday, then I don't get paid. **44. CONVERSE:** If beggars would ride, then wishes were horses. **INVERSE:** If wishes were not horses, then beggars would not ride. **CONTRAPOSITIVE:** If beggars would not ride, then wishes were not horses. **45. CONVERSE:** If you have fewer cavities, then you brush your teeth with Smiles toothpaste. **INVERSE:** If you do not brush your teeth with Smiles toothpaste, then you do not have fewer cavities. **CONTRAPOSITIVE:** If you do not have fewer cavities, then you do not brush your teeth with Smiles toothpaste. **46. CONVERSE:** If your grandchildren will not survive, then we do not stop the pollution of our environment. **INVERSE:** If we stop the pollution of our environment, then your grandchildren will survive. **CONTRAPOSITIVE:** If your grandchildren will survive, then we stop the pollution of our environment. **47.** Let c: You complete this questionnaire; s: You submit this questionnaire; p: Your response to item 34a was "California"; q: Your response to item 37 was "yes." $(c \wedge s)$ only if $[p \vee (\sim q)]$; symbolically, $(c \wedge s) \rightarrow [p \vee (\sim q)]$ **48.** No, they wanted to say "if" instead of "only if."
49. $[p \vee (\sim q)] \rightarrow (c \wedge s)$
50.

p	q	$\sim q$	$p \vee (\sim q)$
T	T	F	T
T	F	T	T
F	T	F	F
F	F	T	T

51. Are you a California resident? If not, did you answer "yes" to item 37? If yes, then complete and submit this residency questionnaire.

Problem Set 4

1. $\triangle ABC$ is a right triangle if and only if $a^2 + b^2 = c^2$. **2.** The judge sentences Otto if and only if the prosecution can prove his guilt. **3.** $x \cdot y = 0$ if and only if $x = 0$ or $y = 0$. **4.** P is a prime if and only if it is a natural number with exactly two divisors. **5.** The United States will develop greater strategic weapons if and only if Russia develops greater strategic weapons. **6.** A triangle is equilateral if and only if it is equiangular. **7.** $p \rightarrow q$ **8.** $\sim p \rightarrow (\sim q)$ **9.** $\sim (p \wedge q)$ **10.** $\sim p \wedge q$ **11.** $[(p \rightarrow q) \wedge p] \rightarrow q$ *Answers to Problems 12-16 may vary.* **12.** If this problem can be solved, then Ron likes this problem. **13.** This problem can be solved because Ron likes this problem. **14.** Neither does Ron like this problem nor can this problem be solved. **15.** Either this problem can be solved or Ron likes this problem. **16.** If this problem can be solved, then Ron likes this problem, and Ron does not like this problem; therefore this problem can't be solved.
17.

p	q	$p \rightarrow q$	$\sim p$	$\sim p \vee q$	$(p \rightarrow q) \leftrightarrow (\sim p \vee q)$
T	T	T	F	T	T
T	F	F	F	F	T
F	T	T	T	T	T
F	F	T	T	T	T

18.

p	q	$\sim q$	$p \to (\sim q)$	$\sim p$	$\sim p \vee (\sim q)$	$[p \to (\sim q)] \leftrightarrow [\sim p \vee (\sim q)]$
T	T	F	F	F	F	T
T	F	T	T	F	T	T
F	T	F	T	T	T	T
F	F	T	T	T	T	T

19. F **20.** F **21.** F **22.**

p	q	$\sim q$	$\sim q \to p$
T	T	F	T
T	F	T	T
F	T	F	T
F	F	T	F

23.

p	q	$\sim p$	$\sim q$	$\sim p \wedge (\sim q)$
T	T	F	F	F
T	F	F	T	F
F	T	T	F	F
F	F	T	T	T

24.

p	q	$p \vee q$	$p \wedge q$	$\sim (p \wedge q)$	$(p \vee q) \wedge [\sim (p \wedge q)]$
T	T	T	T	F	F
T	F	T	F	T	T
F	T	T	F	T	T
F	F	F	F	T	F

25.

p	q	$\sim p$	$\sim q$	$\sim p \vee (\sim q)$
T	T	F	F	F
T	F	F	T	T
F	T	T	F	T
F	F	T	T	T

26. **and:** $(p \wedge q) \leftrightarrow (p|q)|(p|q)$ **or:** $(p \vee q) \leftrightarrow (p|p)|(q|q)$ **not:** $\sim p \leftrightarrow p|p$ *Answers to Problems 27–37 may vary.* **27.** Let p: $x = 10$; q: $2x = 20$. Then $p \leftrightarrow q$. **28.** Let p: $3x + 2 = 11$; q: $x = 3$. Then $p \leftrightarrow q$. **29.** Let s: Smoking is good for your health; d: Drinking is good for your health. Translate *neither s nor d* into symbols: $\sim s \wedge (\sim d)$. **30.** Let h: I will buy a new house; p: All provisions of this sale are clearly understood. Translate $\sim h$ *unless p* into symbols: $\sim p \to (\sim h)$. **31.** Let m: man; i: island. Translate *no m is i* into symbols: $m \to \sim i$. **32.** Let d: A function is differentiable; c: A function is continuous. Then $d \leftrightarrow c$. **33.** Let g: I can go with you; e: I have a previous engagement. Translate $\sim g$ *because e* into symbols: $e \to (\sim g)$. **34.** Let i: I will invest my money in stocks; s: I will put my money in a savings account. Translate *either i or s* into symbols: $(i \vee s) \wedge [\sim (i \wedge s)]$. **35.** Let s: A triangle has two sides equal; i: A triangle is isosceles. Then $s \leftrightarrow i$. **36.** Let h: One who has once heartily laughed; w: One who has once wholly laughed; b: One who is altogether irreclaimably bad. Translate *no* $(h \wedge w)$ *is b* into symbols: $(h \wedge w) \to (\sim b)$. **37.** Let r: Revolution is justified in a moral point of view; f: The majority, by mere force of numbers, deprives a minority of a clearly written constitutional right. Translate *r iff* into symbols: $f \to r$. **38.** c **39.** b and c

Problem Set 5

1. – 8. Answers vary. **9.** law of contraposition **10.** trivial tautology **11.** law of the excluded middle **12.** indirect reasoning **13.** law of double negation **14.** indirect reasoning **15.** law of the excluded middle **16.** law of double negation **17.** law of transitivity
18. direct reasoning **19.** trivial tautology **20.** disjunctive syllogism **21.** indirect reasoning **22.** disjunctive syllogism **23.** law of contraposition **24.** indirect reasoning **25.** disjunctive

syllogism **26.** law of double negation **27.** law of transitivity **28.** law of contraposition **29.** direct reasoning **30.** law of contraposition **31.** law of the excluded middle **32.** trivial tautology **33.** disjunctive syllogism

34.

p	q	r	$p \to q$	$q \to r$	$(p \to q) \wedge (q \to r)$	$p \to r$	$[(p \to q) \wedge (q \to r) \to (p \to r)$
T	T	T	T	T	T	T	T
T	T	F	T	F	F	F	T
T	F	T	F	T	F	T	T
T	F	F	F	T	F	F	T
F	T	T	T	T	T	T	T
F	T	F	T	F	F	T	T
F	F	T	T	T	T	T	T
F	F	F	T	T	T	T	T

All are true, so we can write $\Leftrightarrow$.

35.

p	q	$p \wedge q$	$\sim(p \wedge q)$	$\sim p$	$\sim q$	$\sim p \vee (\sim q)$	$\sim(p \wedge q) \leftrightarrow [\sim p \vee (\sim q)]$
T	T	T	F	F	F	F	T
T	F	F	T	F	T	T	T
F	T	F	T	T	F	T	T
F	F	F	T	T	T	T	T

All are true, so we can write $\Leftrightarrow$.

36.

p	q	$p \vee q$	$\sim(p \vee q)$	$\sim p$	$\sim q$	$\sim p \wedge (\sim q)$	$\sim(p \vee q) \leftrightarrow [\sim p \wedge (\sim q)]$
T	T	T	F	F	F	F	T
T	F	T	F	F	T	F	T
F	T	T	F	T	F	F	T
F	F	F	T	T	T	T	T

All are true, so we can write $\Leftrightarrow$.

37.

p	$\sim p$	$p \wedge \sim p$	$\sim[p \wedge (\sim p)]$
T	F	F	T
F	T	F	T

38.

p	q	$p \to q$	$\sim p$	$\sim p \vee q$	$(p \to q) \leftrightarrow (\sim p \vee q)$
T	T	T	F	T	T
T	F	F	F	F	T
F	T	T	T	T	T
F	F	T	T	T	T

All are true, so we can write $\Leftrightarrow$.

39.

p	$p \wedge p$	$(p \wedge p) \leftrightarrow p$
T	T	T
F	F	T

All are true, so we can write $\Leftrightarrow$.

40.

p	$p \vee p$	$(p \vee p) \leftrightarrow p$
T	T	T
F	F	T

All are true, so we can write $\Leftrightarrow$.

41.

p	q	$p \wedge q$	$p \vee (p \wedge q)$	$[p \vee (p \wedge q)] \leftrightarrow p$
T	T	T	T	T
T	F	F	T	T
F	T	F	F	T
F	F	F	F	T

All are true, so we can write $\Leftrightarrow$.

42.

p	q	$p \vee q$	$p \wedge (p \vee q)$	$[p \wedge (p \vee q)] \leftrightarrow p$
T	T	T	T	T
T	F	T	T	T
F	T	T	F	T
F	F	F	F	T

All are true, so we can write $\Leftrightarrow$.

Problem Set 6

1.

p	q	r	$p \rightarrow q$	$p \rightarrow r$	$(p \rightarrow q) \rightarrow (p \rightarrow r)$	$q \rightarrow r$	$[(p \rightarrow q) \wedge (p \rightarrow r)] \rightarrow (q \rightarrow r)$
T	T	T	T	T	T	T	T
T	T	F	T	F	F	F	T
T	F	T	F	T	T	T	T
T	F	F	F	F	T	T	T
F	T	T	T	T	T	T	T
F	T	F	T	T	T	F	F
F	F	T	T	T	T	T	T
F	F	F	T	T	T	T	T

Since the last column is not all Ts , it is an invalid argument. This is the false chain pattern.

2.

p	q	$p \rightarrow q$	$\sim p$	$(p \rightarrow q) \wedge (\sim p)$	$\sim q$	$[(p \rightarrow q) \wedge (\sim p)] \rightarrow (\sim q)$
T	T	T	F	F	F	T
T	F	F	F	F	T	T
F	T	T	T	T	F	F
F	F	T	T	T	T	T

Fallacy of denying the antecedent.

3. valid **4.** invalid; fallacy of denying the antecedent **5.** invalid; false chain pattern **6.** invalid; fallacy of denying the antecedent **7.** invalid; fallacy of denying the antecedent **8.** valid **9.** valid **10.** valid **11.** valid **12.** invalid; fallacy of assuming the consequent **13.** invalid; fallacy of assuming the consequent **14.** valid **15.** valid **16.** invalid; the converse is not equivalent to the original statement **17.** valid **18.** valid

Problem Set 7

1. T **2.** F **3.** T **4.** T **5.** T **6.** $\exists x$ such that $5x + 7 = 13$ **7.** $\exists y$ such that $3y < 10$ **8.** $\forall x,\ 5x \cdot 1 = 5x$ **9.** $\forall x \forall y,\ xy = yx$ **10.** $\forall a \forall b \forall c,\ a(b + c) = ab + ac$ **11.** $\exists$ **12.** $\forall$ **13.** $\exists$ **14.** $\exists$ **15.** $\forall$

16.

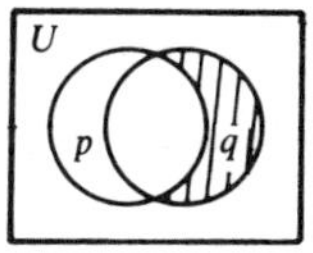

17.

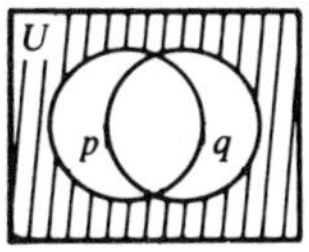

18.

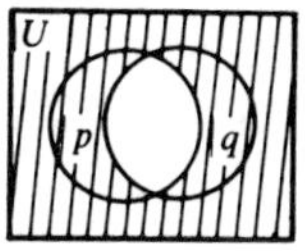

19.

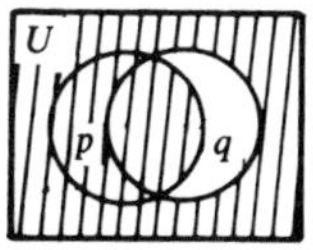

20.

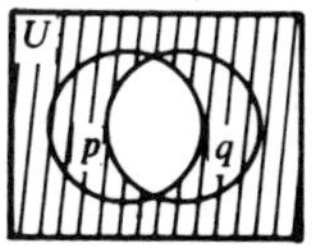

21.

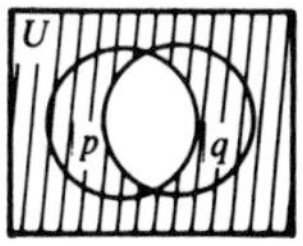

22.

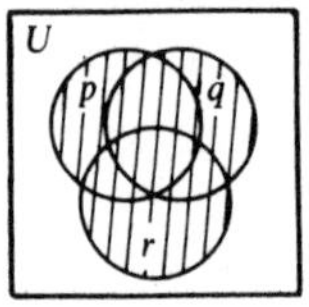

23.

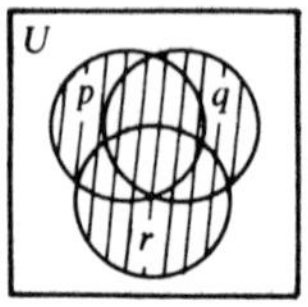

24.

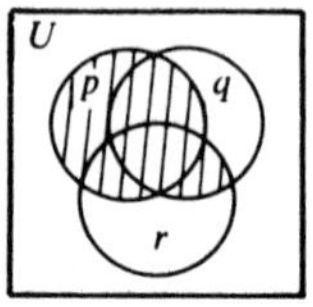

25.

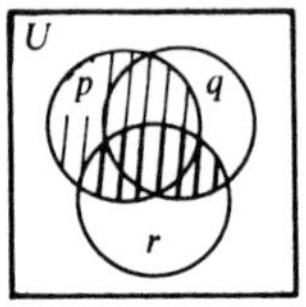

26.

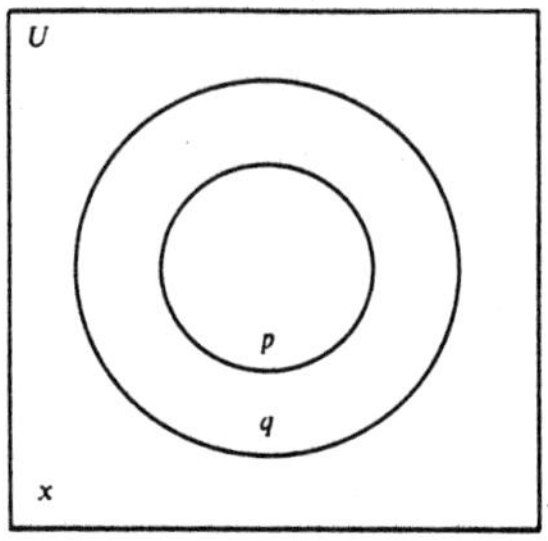

$$\begin{array}{l} p \rightarrow q \\ \sim q \\ \hline \therefore \ \sim p \end{array}$$

27. Let *U*: all people
m: mathematicians
e: eccentric people
r: rich people

Valid

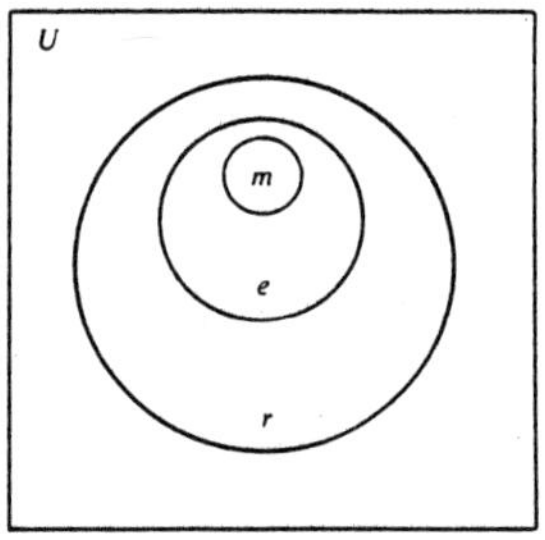

28. Let *U*: all people
w: beautiful women
b: blondes
f: people who have more fun

Valid

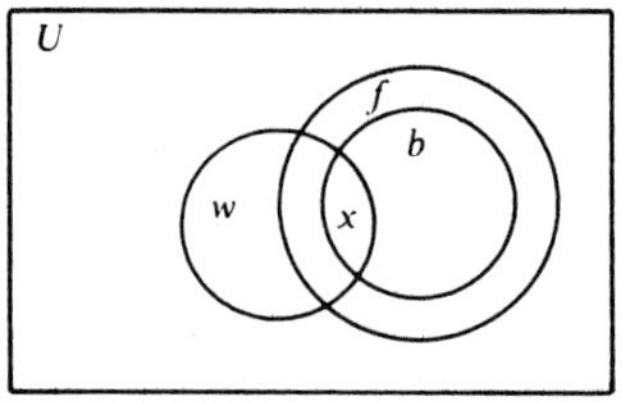

29. Let *U*: all people
b: bachelors
h: handsome people
d: people who drink lemonade

Valid

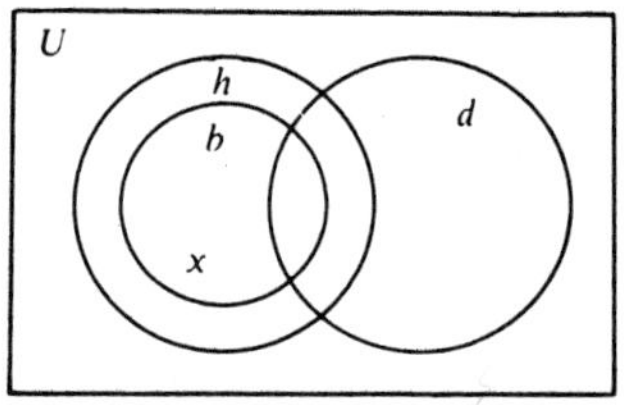

30. Let *U*: all books
d: dictionaries
u: useful books
v: valuable books
Valid

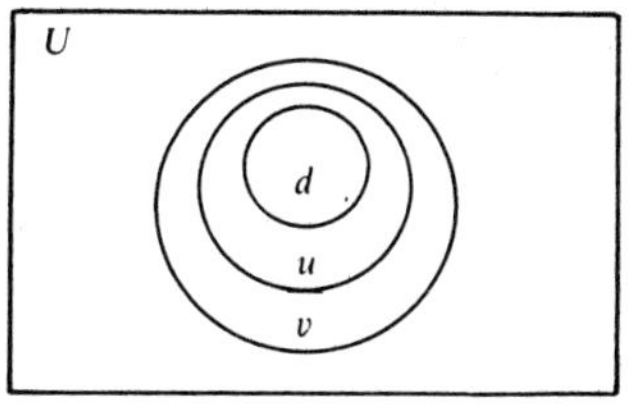

31. Let *U*: all people
p: professors
i: ignorant people
v: vain people
Not valid (see *x*)

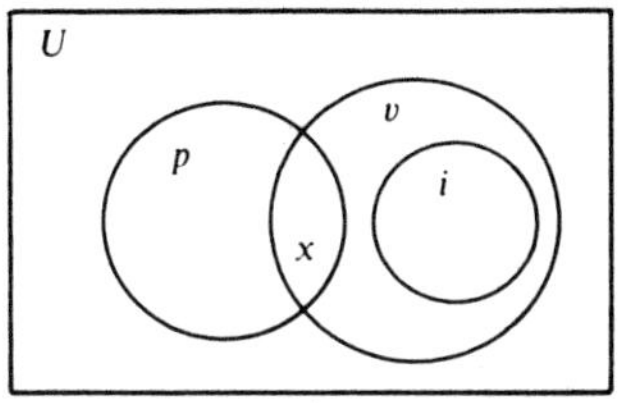

32. Let *U*: all creatures
m: monkeys
s: soldiers
c: mischievous creatures
Valid (provided the set of monkeys is not empty)

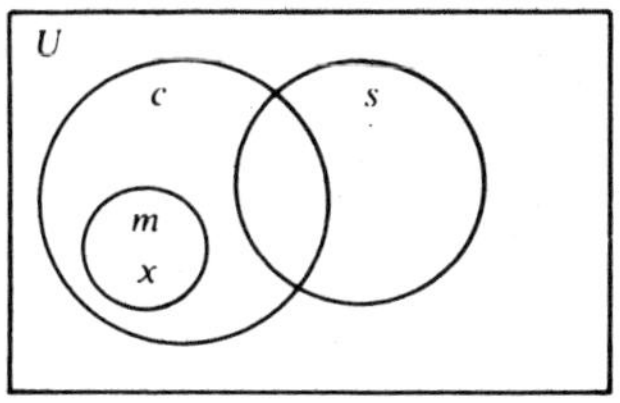

33. Let *U*: all creatures
l: lions
f: fierce creatures
d: coffee drinkers
Not valid

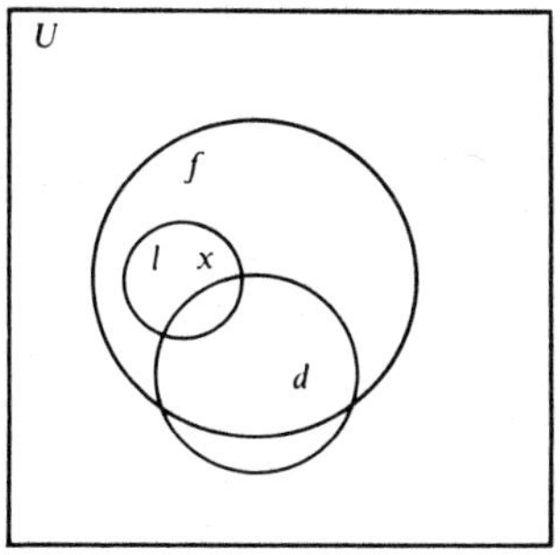

Problem Set 8

The form of the answers may vary. Steps in the reasoning, as well as reasons, should be shown.
1. If you learn mathematics, then you understand women. **2.** I become lazy. **3.** We do not go to the concert. **4.** If you climb the highest mountain, then you are happy. **5.** All Trebbles are expensive. **6.** $b = 0$ **7.** $a = 0$ or $b = 0$ **8.** If I buy my wife a purse, then it cost my father-in-law $5,000. **9.** We do not interfere with the publication of false information. **10.** Two does not divide N or N is not greater than 2. **11.** Babies cannot manage a crocodile. **12.** No hummingbirds are large. **13.** None of my poultry are officers. **14.** No son of yours is fit to serve on a jury. **15.** Airsecond Aircraft Company suffers financial setbacks. **16.** You obey the law. **17.** No kitten with green eyes will play with a gorilla. **18.** These examples are difficult.

Problem Set 9

1. Gary was 17 and had 15 marbles; Harry was 10 and had 12 marbles; Iggy was 3 and had 12 marbles; Jack was 18 and had 9 marbles. They shot in order Gary, Jack, Harry, Iggy. **2.** White, since this could happen only at the north pole. **3.** Curly committed the murder. **4.** Three weighings **5.** Pitcher: Jones; catcher: Smith; first base: Brown; second base: White; third base: Adams; shortstop: Miller; left field: Green; center field: Hunter; right field: Knight.

Problem Set 10

1. Answers vary. **2. a.** battery **b.** light **c.** switch **d.** AND-gate **e.** NOT-gate

3.

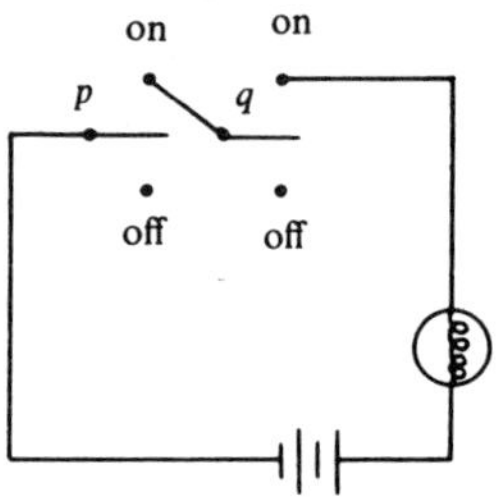

4.

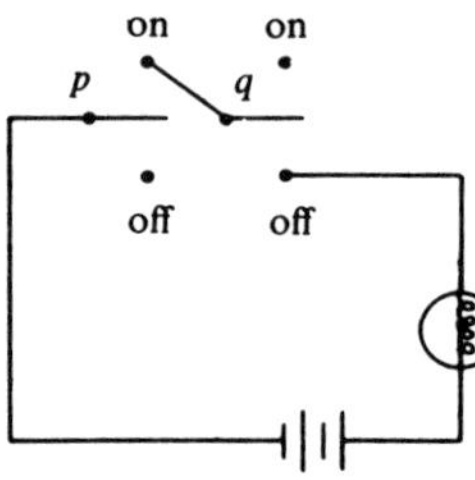

5.

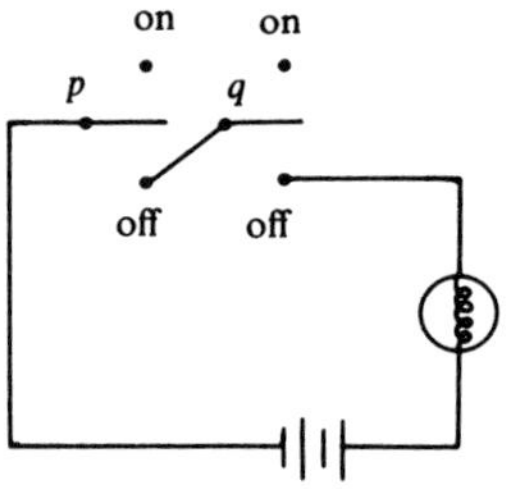

6.

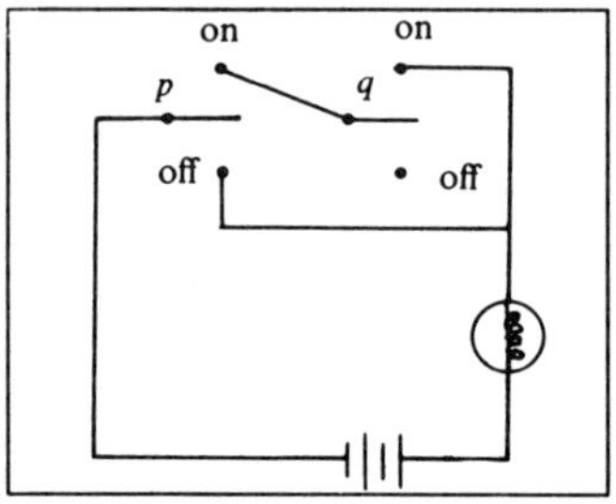

7.

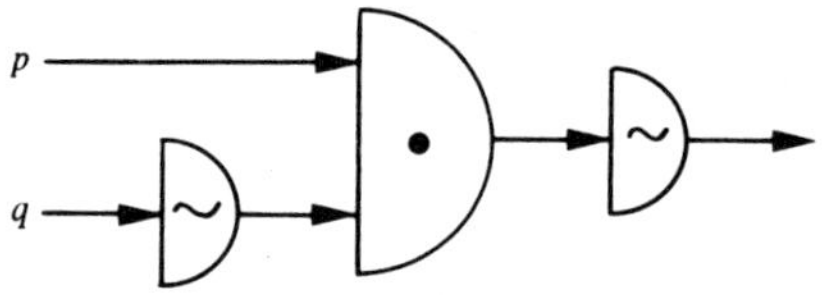

8.

p	q	$p \to q$	$\sim p$	$\sim p \vee q$
T	T	T	F	T
T	F	F	F	F
F	T	T	T	T
F	F	T	T	T

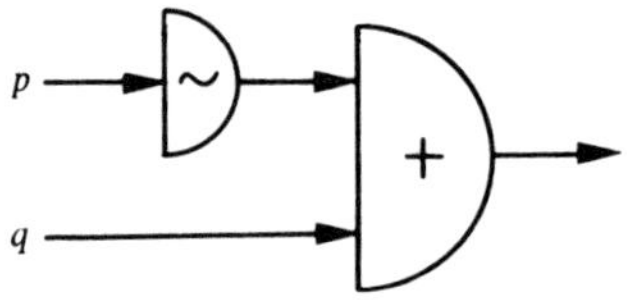

The results of $p \to q$ and $\sim p \vee q$ are the same.

9.

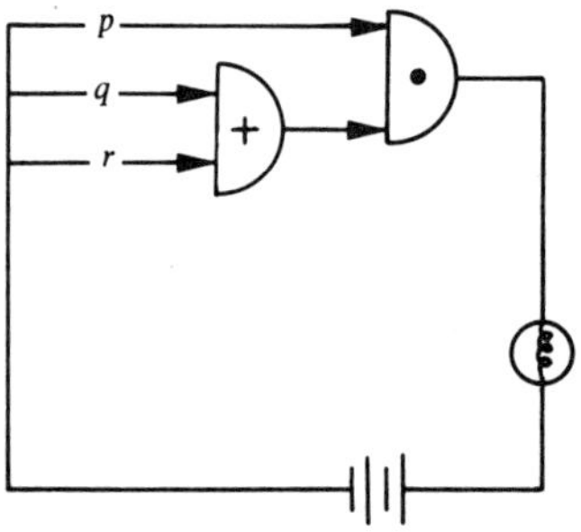

10.

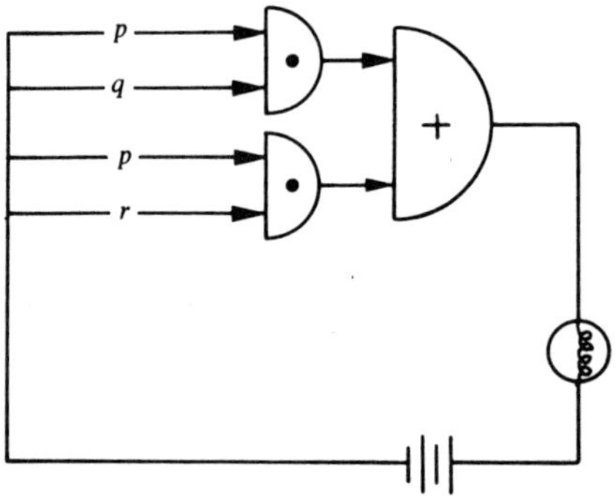

11.

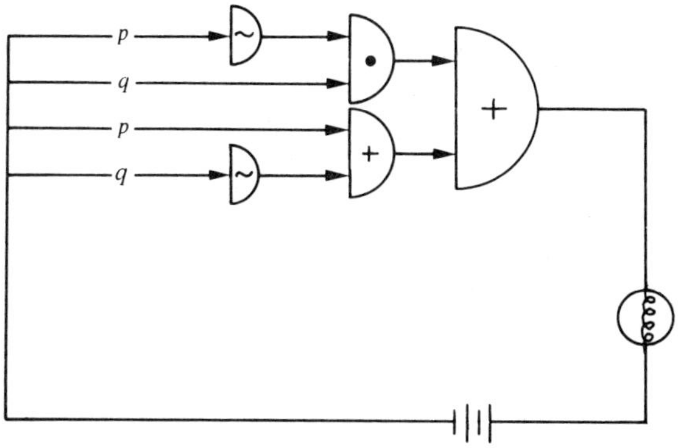

12.

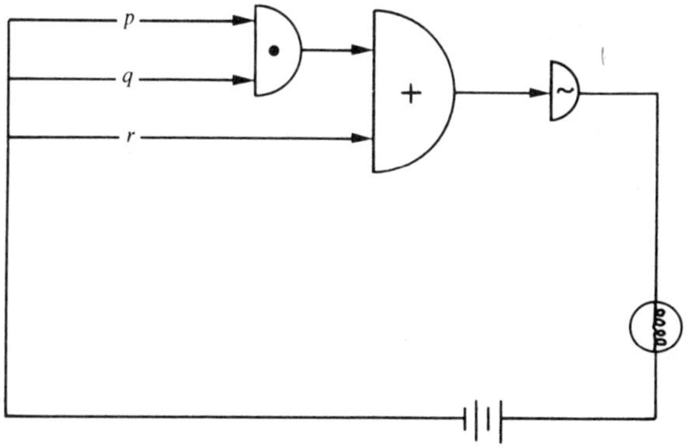

13. **a.**

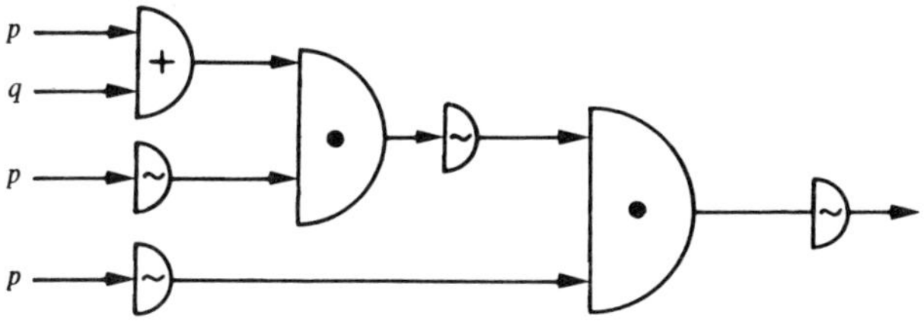

b. \$140,000

c.

p	q	$p \vee q$	$\sim p$	$(p \vee q) \wedge (\sim p)$	$\sim [(p \vee q) \wedge (\sim p)]$	$\sim [(p \vee q) \wedge (\sim p)] \wedge (\sim p)$	ANS
T	T	T	F	F	T	F	T
T	F	T	F	F	T	F	T
F	T	T	T	T	F	F	T
F	F	F	T	F	T	T	F

They are the same.

d.

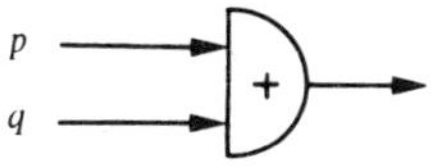

e. cost, \$20,000; savings, \$120,000

Note: Problems 14–21 can also be solved by designing the circuits using AND-, OR-, and NOT-gates.

14.

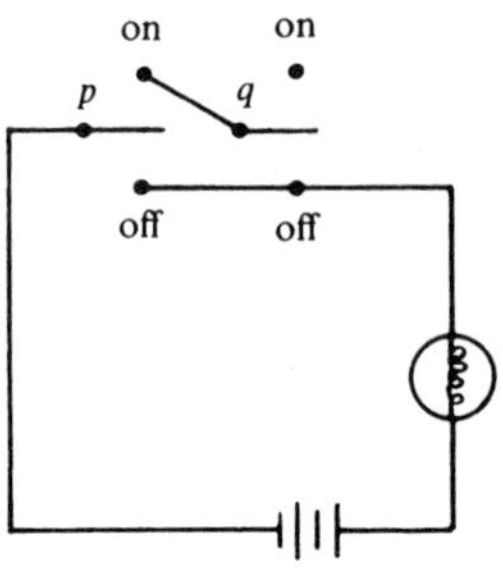

15.

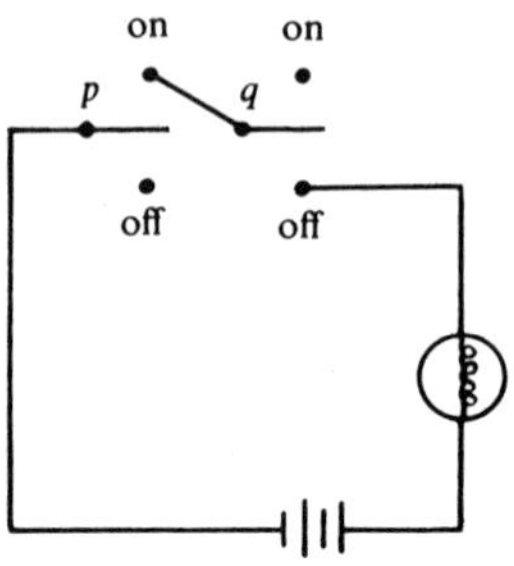

16.

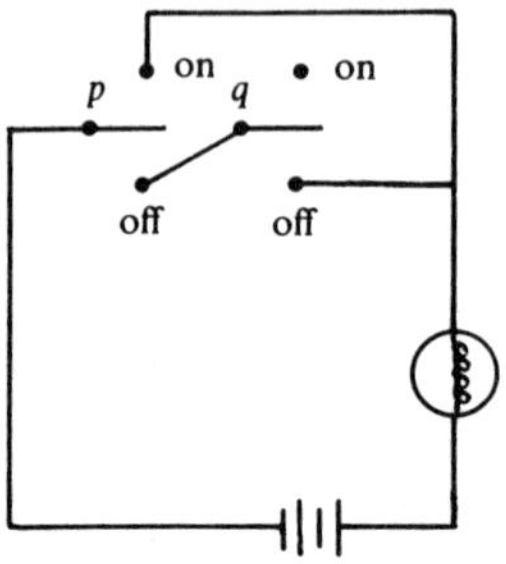

17. Same as 15 **18.** Same as 6 **19.**

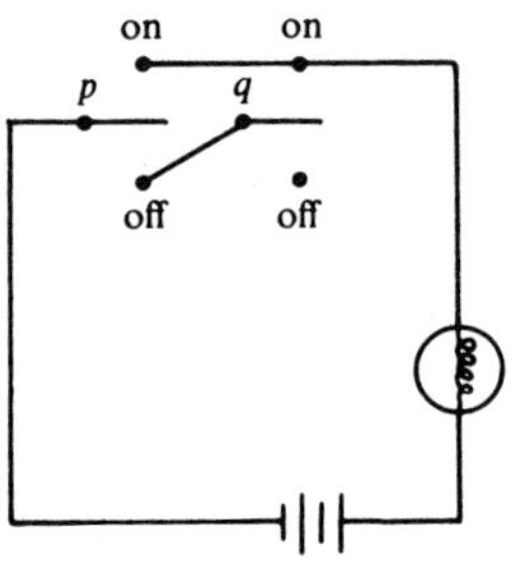

20. Same as 14

21.

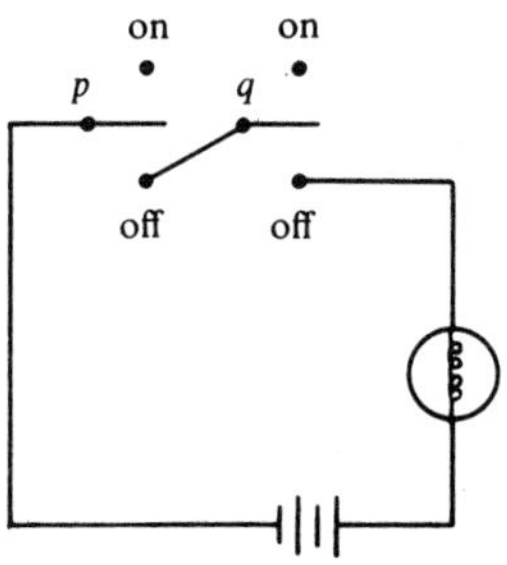

22. Let the committee members be *a*, *b*, and *c*, respectively. Light *on* represents a majority *yea* vote.

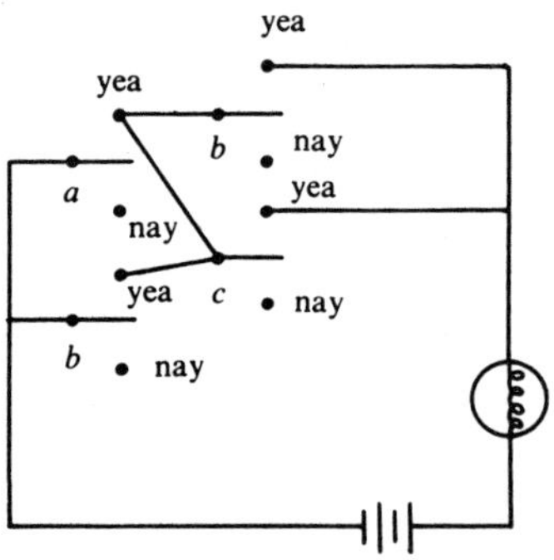

23. Let the committee members be *a*, *b*, *c*, *d*, and *e*. Light on represents a majority *yea* vote.

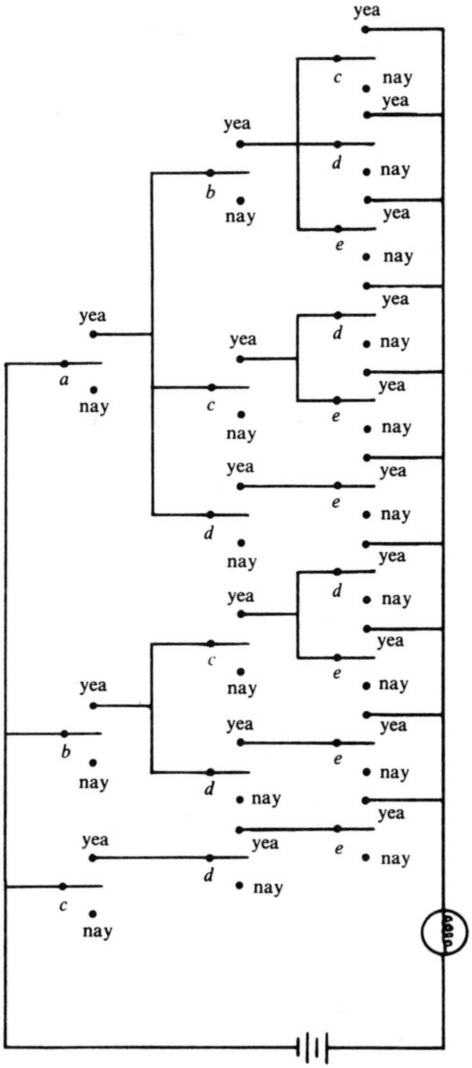

Problem Set 11

1.

p	q	$p \vee q$	$q \vee p$	$(p \vee q) \leftrightarrow (q \vee p)$
T	T	T	T	T
T	F	T	T	T
F	T	T	T	T
F	F	F	F	T

Always true, so we can write $\Leftrightarrow$.

2.

p	q	r	$p \vee q$	$(p \vee q) \vee r$	$q \vee r$	$p \vee (q \vee r)$	$[(p \vee q) \vee r] \leftrightarrow [p \vee (q \vee r)]$
T	T	T	T	T	T	T	T
T	T	F	T	T	T	T	T
T	F	T	T	T	T	T	T
T	F	F	T	T	F	T	T
F	T	T	T	T	T	T	T
F	T	F	T	T	T	T	T
F	F	T	F	T	T	T	T
F	F	F	F	F	F	F	T

Always true, so we can write $\Leftrightarrow$.

3. Yes. Answer is supported by constructing the truth table for $[p \vee (q \wedge r)] \leftrightarrow [(p \vee q) \wedge (p \vee r)]$. **4.** No. Answer is supported by constructing the truth table for $[\sim (p \wedge q)] \leftrightarrow [\sim p \wedge \sim q]$ and showing that the result is not a tautology. **5.** For any proposition p and operation $*$, there exists a unique element i such that $p * i = i * p = p$. **6.** A Boolean algebra is a mathematical system with two operations satisfying the closure, commutative, associative, distributive, identity, and complementation properties for both operations. **7.** Yes; result should be verified.

Problem Set 12

1. Definitions are established by agreement, and theorems are results proved by using definitions, axioms, previous theorems, and the rules of logic.

2.

p	q	$\sim p$	$p \wedge q$	$p \vee q$	$p \rightarrow q$	$p \leftrightarrow q$
T	T	F	T	T	T	T
T	F	F	F	T	F	F
F	T	T	F	T	T	F
F	F	T	F	F	T	T

3.

p	q	$p \wedge q$	$\sim (p \wedge q)$
T	T	T	F
T	F	F	T
F	T	F	T
F	F	F	T

4.

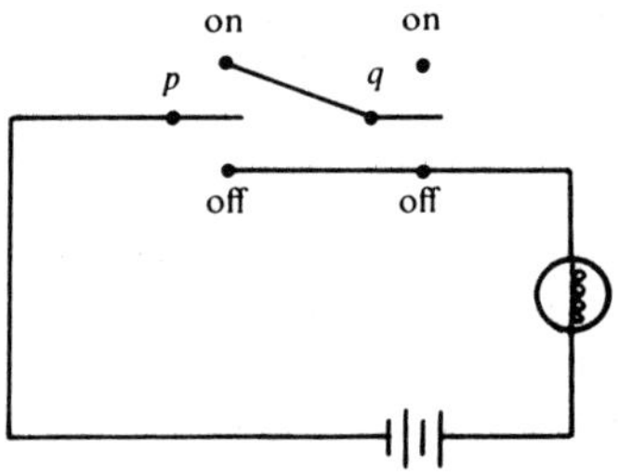

5.

p	q	$p \wedge q$	$\sim (p \wedge q)$	$\sim p$	$\sim q$	$\sim p \vee (\sim q)$	$\sim (p \wedge q) \leftrightarrow [(\sim p) \vee (\sim q)]$
T	T	T	F	F	F	F	T
T	F	F	T	F	T	T	T
F	T	F	T	T	F	T	T
F	F	F	T	T	T	T	T

6.

p	q	$\sim p$	$\sim p \vee q$	$(\sim p \vee q) \wedge p$	$\sim [(\sim p \vee q) \wedge p]$
T	T	F	T	T	F
T	F	F	F	F	T
F	T	T	T	F	T
F	F	T	T	F	T

7.

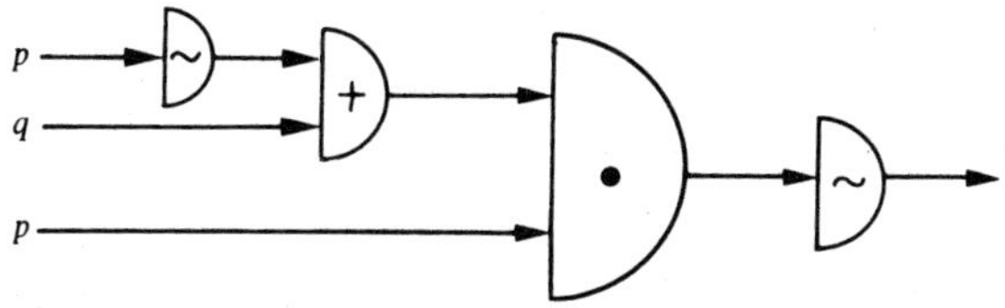

or (compare with Problem 3)

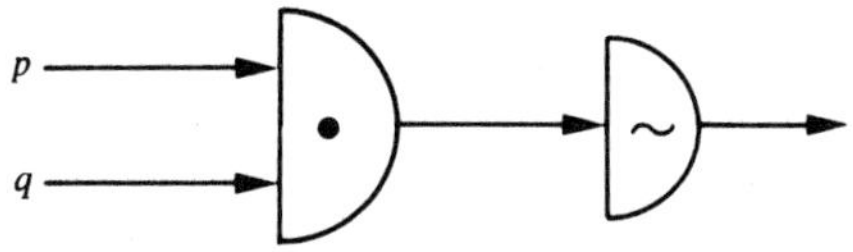

8. $p \rightarrow q$
$\underline{p}$
$\therefore q$

p	q	$p \rightarrow q$	$(p \rightarrow q) \wedge p$	$[(p \rightarrow q) \wedge p] \rightarrow q$
T	T	T	T	T
T	F	F	F	T
F	T	T	F	T
F	F	T	F	T

9. Answers vary. **10.** Answers vary. **11. a.** If P is a prime number, then $P + 2$ is a prime number. **b.** Either P is a prime number or $P + 2$ is a prime number, but not both. **12.** Let c: x is a computer; d: x is capable of self-direction. **a.** $\forall x,\ c \rightarrow (\sim d)$ **b. Contrapositive:** $\forall x,\ d \rightarrow\ \sim c$, or "If it is capable of self-direction, then it is not a computer." **13. Converse:** If Rosamond is not liberated, then she doesn't go to college. **Inverse:** If Rosamond goes to college, then she is not liberated. **Contrapositive:** If Rosamond is liberated, then she goes to college. **14. a.** F **b.** T **15.** yes; indirect reasoning **16.** not valid; fallacy of denying the antecedent **17.** I do not attend to my duties. **18.** If Jack and Jill go up the hill, then it will cost Jack a lot of money.

19.

p	q	r	$p \vee q$	$(p \vee q) \vee r$	$q \vee r$	$p \vee (q \vee r)$	$[(p \vee q) \vee r] \leftrightarrow [p \vee (q \vee r)]$
T	T	T	T	T	T	T	T
T	T	F	T	T	T	T	T
T	F	T	T	T	T	T	T
T	F	F	T	T	F	T	T
F	T	T	T	T	T	T	T
F	T	F	T	T	T	T	T
F	F	T	F	T	T	T	T
F	F	F	F	F	F	F	T

All results are true, so we can write $\Leftrightarrow$.

20. a. $\forall x,\ 5x + 7x = 12x$ **b.** $\exists y$ such that $15y < 10$ **c.** $\exists z$ such that $z + 2 = 5$ **d.** $\forall x \exists y$ such that $x + y = 7$ **e.** $\forall a \forall b,\ a + b = b + a$ **21. a.** A Boolean algebra is a mathematical system with two operations satisfying the closure, commutative, associative, distributive, identity, and complementation properties for both operations. **b.** Answers vary.

22. Let U: all creatures
d: dogs
f: creatures that have fleas

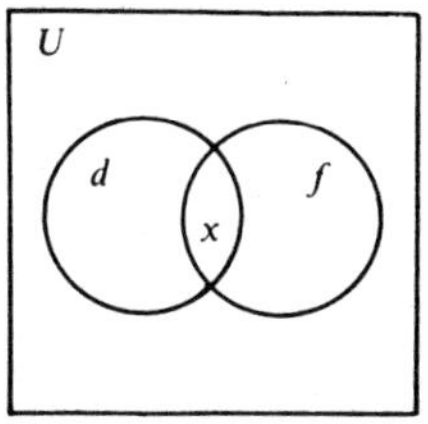

23. **a.** existential **b.** $\exists$ **24.** Answers vary.

25. Let U: all people
d: dancers
e: eccentric people
r: rich people
b: boring people
s: singers
Not valid

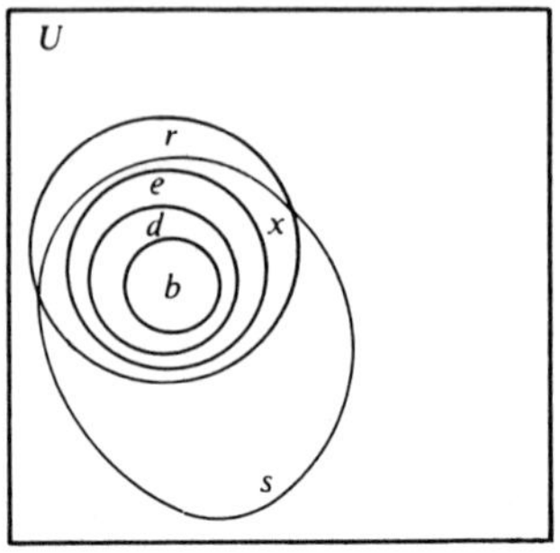

26. Yes; answers vary. See diagram from Problem 25 for other conclusions. **27.** not valid; fallacy of assuming the consequent **28.** not valid; fallacy of assuming the consequent **29.** valid; direct reasoning **30.** valid; direct reasoning **31.** not valid; fallacy of assuming the consequent **32.** All parts are true. **33.** $ad = bc$ if and only if $a/b = c/d$

INDEX

All, negation of, 7
Analysis of statements, 67
And, truth table, 11
Answers, 78
Antecedent, 15
Arguments, 46
Associative laws, Boolean algebra, 71
Associative property, 68, 70
Assuming the consequent, fallacy of, 36
Axiom, 67

Baseball problem, 59
Battery, 61
Bear problem, 58
Biconditional, 23
    definition, 23
    translations, 23
Boolean algebra, 70

Carroll, Lewis, 46
Circuits, 60
    Boolean algebra, 70
    conjunction, 61
    disjunction, 61
    negation, 61
    parallel, 63
    related to logic and sets, 71
    series, 62
Closure property, 67, 70
Coin problem 59
Commutative laws, Boolean algebra, 71
Commutative property, 68, 70
Complementation, property of 69, 70
Complement laws, Boolean algebra, 72
Compound statement, 3
    truth value, 3
Conclusion, 15, 31
Conditional, 15
    definition, 16
    variations, 18
Conjunction, 3
    circuit, 61, 62
    definition, 4
Connective, simple, 44
Consequent, 15
Contradiction, 28
Contrapositive, 18
    law of, 20, 29, 33
    truth table, 19
Converse, 18
    fallacy of, 36
    truth table, 19
Counterexample, 36

Denying the antecedent, fallacy of, 36
Direct reasoning, 29
    proof, 30, 31
Disjunction, 4
    circuit, 61
    definition, 5
    syllogism, 29, 33
Distributive laws, Boolean algebra, 71
Distributive property, 68, 70
Double negation, law of, 29
    proof, 30

Either or,
    definition, 24
    truth table, 25
Equivalence, 24
Equivalent, 24
Euler, Leonhard, 42
Euler circles, 41, 42
Excluded middle, law of, proof, 30
Exclusive or, 6
Existential quantifier, 41, 42

Fallacies, 36
    assuming the consequent, 36, 38
    denying the antecedent, 36, 38
    of the converse, 36
    of the inverse, 36
    false chain pattern, 38
False sentence, 41

Hypothesis, 15, 31

Idempotent laws, Boolean algebra, 72
Identity laws, Boolean algebra, 72
Identity property, 69, 70
If and only if, 23
Implication, 29
Inclusive or, 6
Indirect reasoning, 29, 31
    proof, 32
Inverse property, 69
    truth table, 19

Leibniz, Gottfried, 1
Liar problem 54
Light switch, 61
Logical arguments, 46
Logical equivalence, 29
Logical gate, 61, 63
Logic,
    Boolean algebra, 70
    circuits, 60
    laws of, 29
    related to sets and circuits, 71

Marble player problem, 58
Mathematical system, 67

Necessary condition, 23
Negation, 6
    all, 7
    circuit, 61
    definition, 6
    no, 7
    some, 7
Neither nor, 3
    definition, 24
    truth table, 25
No p is q,
    definition, 24
    negation of, 7
    truth table, 25
Not, truth table, 11

Odd ball problem, 56
Off-state, 60
On-state, 60
Open sentence, 41
Operations, 67
  miscellaneous, 24
Or,
  exclusive, 6
  inclusive, 6
  truth table, 11
Outline of book, 73
Overlapping propositions, 44

p because q,
  definition, 24
  truth table, 25
p unless q,
  definition, 24
  truth table, 25
Parallel circuit, 63
Premise, 31
Prisoner problem, 55
Proof, 50
Puzzle problems, 54

Quantifiers, 41

Relation, 67
Reporter's error problem, 56
Review, 73
Richard III, 32
Russell, Bertrand, 1

Sentence,
  false, 41
  open, 41
  true, 41
Series circuit, 62
Sets, 67
  Boolean algebra, 70
  related to logic and
    circuits, 71
Simple connectives, 44
Simple statement, 3
  truth value, 3
Some, negation of, 7
Statements, 2
  analysis of, 67
  combining, 8
  compound, 3
  simple, 3
Sufficient condition, 23
Summary, 73
Switches, 60
Symbolic logic, 1
System, mathematical, 67

Tautology, 28
Transitivity, law of, 29, 32
Trivial tautology, 29
  proof, 30
True sentence, 41
Truth table, 11
Truth value, 3
  with universal quantifier, 42
Turkey problem, 58

Universal characteristic, 1
Universal quantifier, 41
Unless, 3

Valid argument, 36
Validity, 36
Venn, John, 42

Whitehead, Alfred North, 1
Whodunit problem, 58